YEARS 5 & 6

MULTIPLICATION & DIVISION

Ann Baker

Illustrated by
Janice Bowles

About this book

Each unit in this book begins with a brief **explanation** of a concept or a strategy. You are encouraged to read this explanation with your child and, where appropriate, to use everyday materials and examples to give meaning to the concepts.

We practise is a worked example for you and your child to discuss together, paying particular attention to the thinking processes required to understand the concept or apply the strategy.

You practise gives your child the opportunity to practise the concept or strategy. It also indicates how well your child understands the new material and often includes problem-solving questions to ensure that your child has mastered the concept or strategy.

If further support is required, you and your child's teacher can devise a plan to ensure that all the basic concepts are fully understood and consolidated.

The **Tests** at the end of the book are provided to check that the concepts are fully understood. Test 1 can be done after units 1–10 are completed and Test 2 when the book is finished.

Meet 'BOB' – Back Of the Book

At the end of each unit, BOB reminds your child to go to the Answers section at the back of the book.

Mathematical Content

This book has been designed to cover the concepts of multiplication and division that your child will encounter in **Year 5** and **Year 6**. The units provide a comprehensive coverage of the following Key Topics from the **Australian Curriculum: Mathematics.**

Australian Curriculum : Mathematics

YEAR 5

Solve problems involving multiplication of large numbers by one- or two-digit numbers using efficient mental, written strategies and appropriate digital technologies (ACMNA100)

Solve problems involving division by a one digit number, including those that result in a remainder (ACMNA101)

Use equivalent number sentences involving multiplication and division to find unknown quantities (ACMNA121)

YEAR 6

Select and apply efficient mental and written strategies and appropriate digital technologies to solve problems involving all four operations with whole numbers (ACMNA123)

Multiply decimals by whole numbers and perform divisions that result in terminating decimals, with and without digital technologies (ACMNA129)

Multiply and divide decimals by powers of 10 (ACMNA130)

Contents & Checklist

WRITING and TALKING ABOUT MULTIPLICATION and DIVISION

Multiplication

There are many words that mean multiply.

For example, **5 × 4** can be said as:

5 **times** 4

5 **multiplied by** 4

5 **lots of** 4

5 **equal-sized groups of** 4

The result of a multiplication is called the **product**. For example, in 5 × 4 = 20, 20 is called the product of 5 and 4.

The other parts of a multiplication are:

5 × 4 = 20

multiplier (5), multiplicand (4), product (20)

Division

There are many words that mean divide.

For example **20 ÷ 4 = 5** can be said as:

20 **divided by** 4 equals 5

20 **can be shared into** five groups of 4

4 **goes into** 20 five times

The result of a division is called the **quotient**.

The other parts of a division are:

20 ÷ 4 = 5

dividend (20), divisor (4), quotient (5)

When multiplying, it doesn't matter if you swap the numbers around.

4 × 5 is the same as 5 × 4.

Fact family

Multiplication and division are related in the same way as addition and subtraction.

3 × 4 = 12	4 × 3 = 12
12 ÷ 3 = 4	12 ÷ 4 = 3

All four facts are part of **one multiplication fact family** and you will often find that knowing a multiplication fact helps you find the answer to a division.

GAME CARD IDEAS

Cut out the Game Cards and the spinner. Pierce a small hole in the centre of the spinner and push a matchstick through the hole to make the spinner's handle. Both the cards and spinner will last longer if they are laminated.

MULTIPLES

A game for two or more players.

Use all the cards and the spinner for this game. Each player is dealt six cards. The player with the highest number on a single card spins the spinner first. On their turn, if a player has any card with a number that is a multiple of the number that the spinner lands on, then this card or cards can be discarded. If they don't, then they have to pick up another card. Players can ask for help or all players can play with their cards showing.

The winner is the first person to discard all their cards.

To make it easier, start by playing with numbers less than 70. When the rules of divisibility (covered in Units 12 and 13) have been understood, then all the number cards can be included.

To settle disputes, use a calculator to check that there is no remainder when the number on the card is divided by the spinner number.

Happy Multiples

A game for 2, 3 or 5 players.

This game is based on the old favourite *Happy Families*, where the aim is to collect all the members of one particular family.
A family consists of three numbers that are multiples of
2, 3, 4, 5 6 or 9.

Deal all the cards evenly to the players. The player with the highest number on a single card goes first. This player chooses a number (from one of their cards) and any player if they have a card that is a multiple of this number. For example, if they had 22 in their hand, they could ask for another multiple of 2. If another player has a card that matches the request, then they have to hand over that card. If no one has a matching card, then it is the next player's turn.

When a player has collected a family, they place it in front of them. The winner is the person who collects the most families.

Note: Make sure players say *please* and *thank you* – being polite is part of the traditional *Happy Families* game and can add an entertaining twist.

NOTE: These games are meant to be fun and provide practice without stress. It is recommended that you stop playing while you are still having fun and then your child will want to play again another time.

DOUBLES and HALVES

Doubling and halving are good strategies to use when you want to multiply or divide mentally (in your head).

You probably know all the doubles of single-digit numbers, but when you **double** a larger number it often helps to **split** it into two parts.

For example, to find **2 × 28**, think of 28 in **two parts**, **20** and **8**. Then you can **double each part**.

× 2: 28 → 20 and 8 → 40 and 16 → 56

To **halve** a number like **36**, number splitting can also help. Think of **36** in **two parts**, **30** and **6**. Then you can **halve each part**.

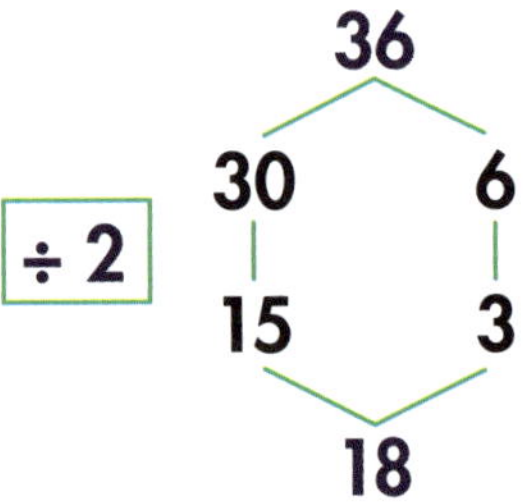

But if you aren't sure what **30 ÷ 2** is, you could use **different number splits** for 36.

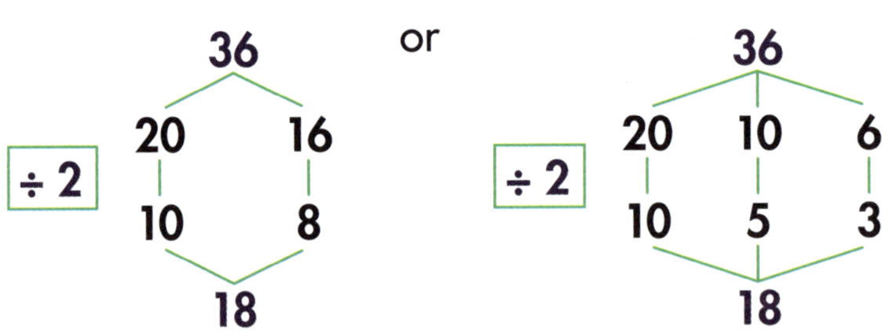

We practise

Use number splitting to double 47.

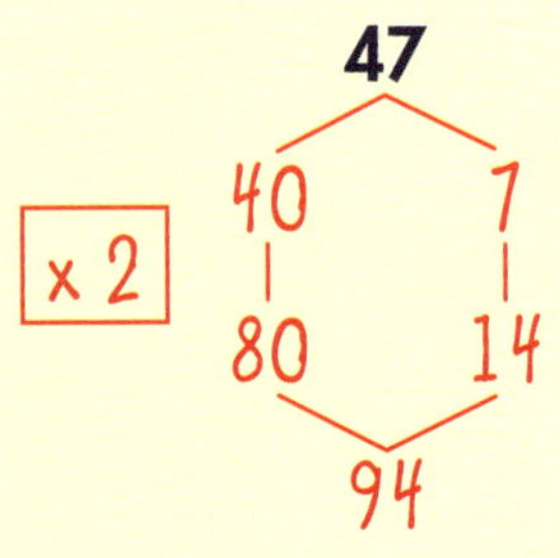

Use number splitting to halve 74.

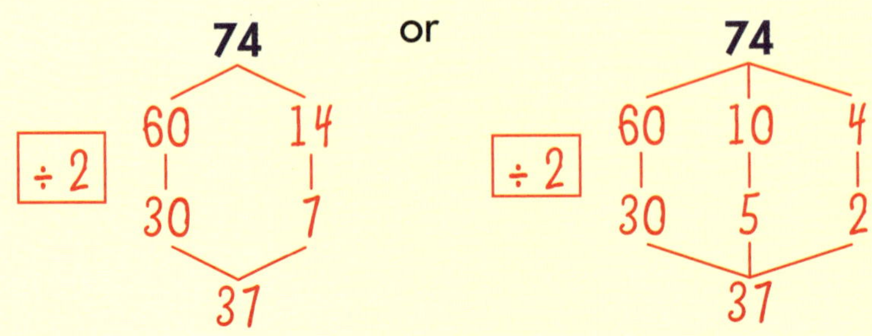

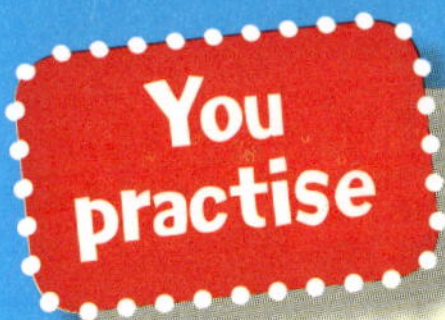

Use number splitting to double these numbers.

34

46

49

58

67

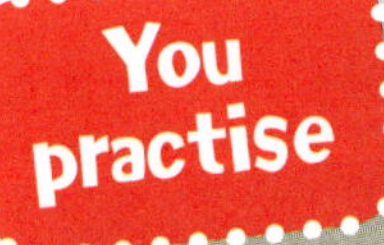

Use number splitting to halve these numbers.

28

34

48

52

74

For a number like 52, where the 10s digit is odd, it might be best to split it into 40 + 12.

MULTIPLYING BY 10 and 100

10 is a friendly number, which means that it is easy to multiply by. Look at these multiplications. Can you see a pattern that works for multiplying all whole numbers?

6 × 10 = 60

60 × 10 = 6 × 10 × 10 = 6 × 100 (60 is 6 × 10)
= **600**

600 × 10 = 6 × 100 × 10 = 6 × 1000 (600 is 6 × 100)
= **6000**

This works for larger numbers too.
16 × 10 = 160

When you multiply by 10, you just put one zero on the end of the number being multiplied.

However, you are not just adding a zero when you multiply by 10. Look at this place diagram to see what happens when you **multiply** 36 by 10.

	100s	10s	1s	
		3	6	× 10
=	3	6	0	

When you multiply by 100, you put two zeroes on the end of the number being multiplied – each digit moves up two places. So 36 × 100 = 3600.

Sometimes both numbers to be multiplied end in zero. When this happens, you can gather the zeroes together for the answer.
Look at this example.

30 × 60 = 3 × 10 × 6 × 10
= 3 × 6 × 10 × 10
= 18 × 100
= **1800**

Complete these multiplications.

34 × 10 = 340

78 × 10 = 780

107 × 10 = 1070

83 × 100 = 8300

30 × 20 = 600

5 × 40 = 200

70 × 80 = 5600

200 × 30 = 6000

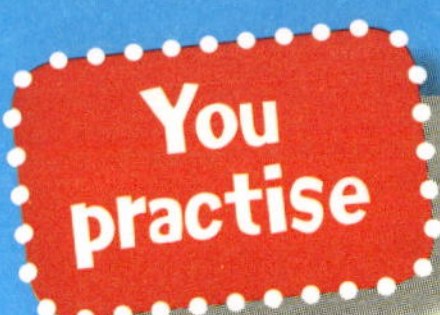

Complete these multiplications.

5 × 10 = __________

35 × 10 = __________

108 × 10 = __________

10 × 56 = __________

85 × 100 = __________

30 × 50 = __________

80 × 70 = __________

140 × 20 = __________

300 × 600 = __________

400 × 700 = __________

Remember to check that you have the right number of zeroes.

UNIT 3

STRATEGY for MENTAL MULTIPLICATION

There is a strategy for working out a multiplication without pencil and paper. The strategy is called number splitting and it can make multiplications easy to do in your head.

To explain how to use number splitting, you need to know these **multiplication words**:

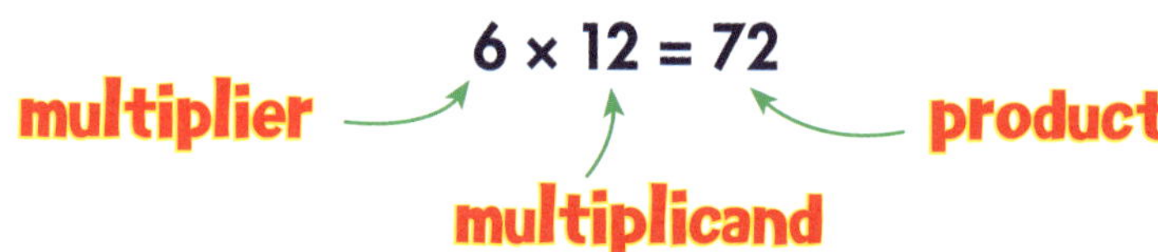

To work out **6 × 12**, you can **split** the **multiplicand** into two parts that are easy to multiply. **12** can be split into **10** and **2**.

Then you can **multiply 10** by **6** and **2** by **6** and add the results.

6 × 12 ⟨ **6 × 10 = 60** / **6 × 2 = 12** ⟩ **72**

You could also split the **multiplier** into its **factors** to make it easier.

6 can be split into **2 × 3**.
So then **6 × 12**, becomes **2 × 3 × 12**.

To work this out, first multiply 3 × 12 = 36, and then multiply 2 × 36 = 72, which is a double.

I can do 3 × 12 in my head. It makes 36.

We practise

Split the multiplicand to find 4 ×14.

4 x 14 ⟨ 4 x 10 = 40 / 4 x 4 = 16 ⟩ 56

Split the multiplier to find 4 × 14.

4 x 14 = 2 x 2 x 14

2 x 14 = 28

2 x 28 = 56

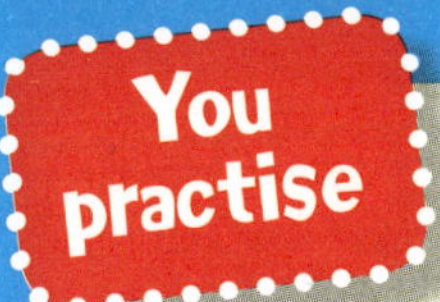

Split the multiplicand for these multiplications.

 3 × 15

See if you can do these in your head. Then check with paper and pencil if you need to.

 4 × 32

 6 × 52

3 5 × 43

 7 × 23

Split the multiplier for these multiplications.

 6 × 14

 8 × 15

 4 × 29

 8 × 73

 6 × 43

BOB time!

THE CHUNKING METHOD

Sometimes a multiplication is just too hard to do it in your head. It's time to use paper and pencil to make sure there are no mistakes.

You have probably used **chunking** when doing two-digit **additions**, such as **36 + 47**. The two numbers are added in chunks – the **30 and 40** followed by the **6 and 7**.

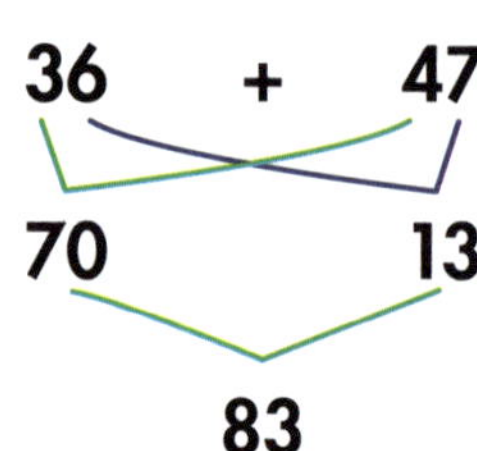

The same method can be used with **multiplication**.

For example, for **36 × 8**, you can split 36 into 30 + 6 and then multiply the chunks by 8.

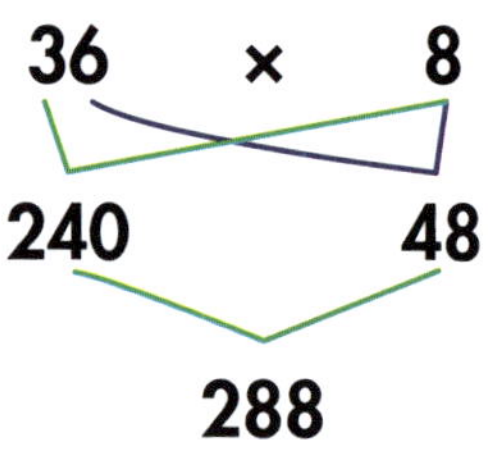

If the **addition** of the **two parts** gets complicated, you can always **add another step** to make sure you are confident at every stage of the process.

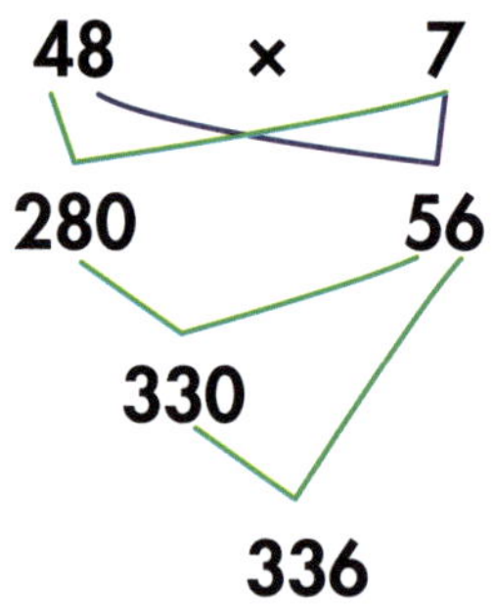

If you are not sure, check your answer with a calculator.

Better to be safe than sorry!

We practise

Show how to use chunking for this multiplication:

43 × 8

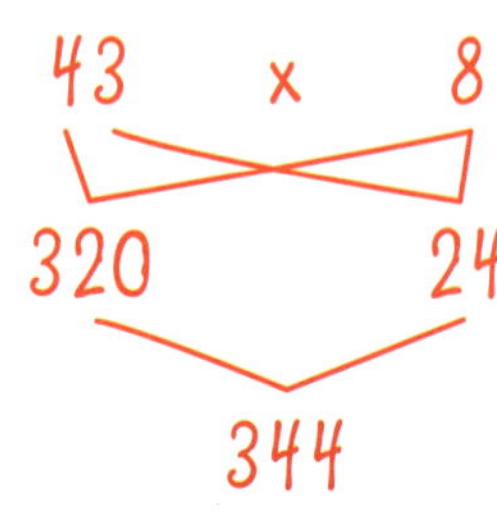

You practise

Use chunking for these multiplications.

1. 23 × 9

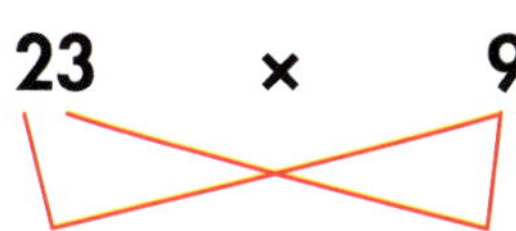

2. 37 × 4

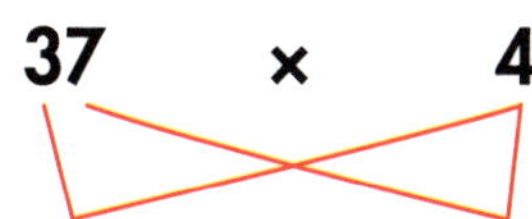

3. 36 × 6

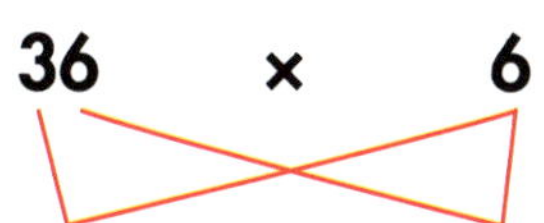

4. 53 × 7

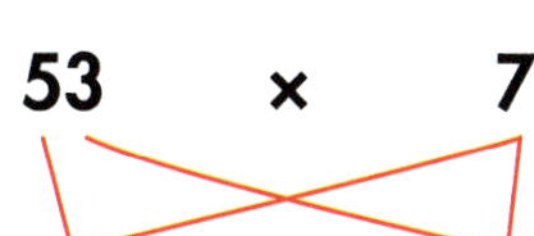

5. 72 × 6

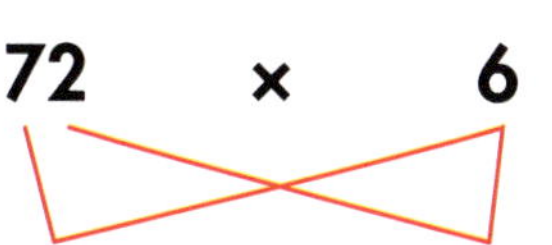

6. 83 × 9

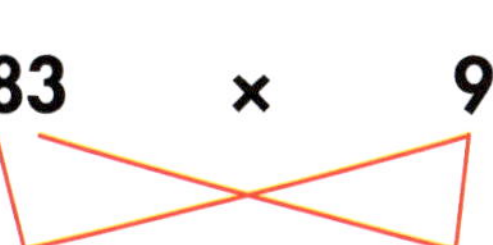

7. 73 × 6

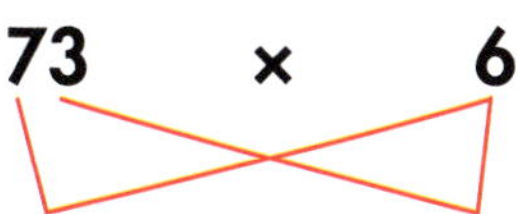

8. 8 × 57

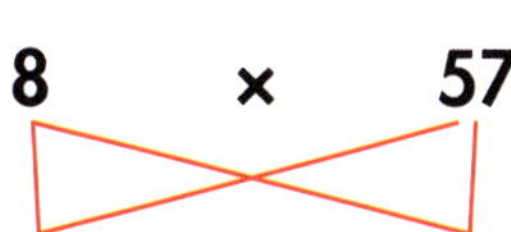

9. 6 × 210

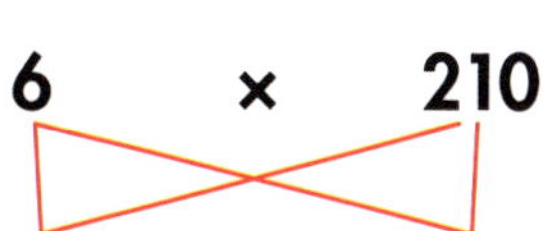

10. 9 × 470

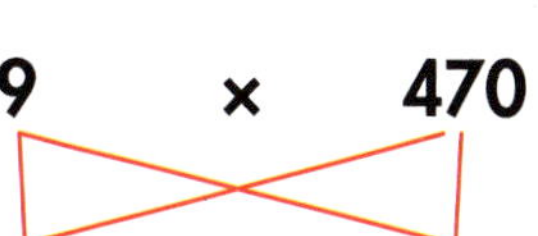

BOB time!

UNIT 5

THE FRONT-END METHOD

When you use the front-end method for multiplication, you can see what the answer will be roughly before you work out the correct answer.

This will help you make sure that you are close to the right answer.

When you use the front-end method, you multiply the two parts of the digits in the multiplication separately.

Look at this example.

```
   3 6
×    8
 2 4 0    30 × 8 = 240
   4 8    6 × 8 = 48
 2 8 8
```

The front-end method can also be used with larger numbers.

Look at this example.

```
    3 6 5
×       5
  1 5 0 0    300 × 5 = 1500
    3 0 0    60 × 5 = 300
      2 5    5 × 5 = 25
  1 8 2 5
```

Use the front-end method to find 47 × 6.

```
   4 7
×    6
 2 4 0    6 × 40 = 240
   4 2    6 × 7 = 42
 2 8 2
```

Use the front-end method to find 326 × 7.

```
   3 2 6
×      7
 2 1 0 0    7 × 300 = 2100
   1 4 0    7 × 20 = 140
     4 2    7 × 6 = 42
 2 2 8 2
```

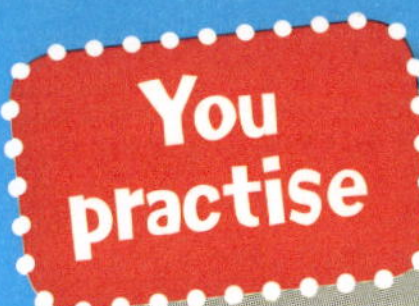

Use the front-end method for these multiplications.

1.
$$\begin{array}{r} 23 \\ \times \quad 4 \\ \hline \end{array}$$

2.
$$\begin{array}{r} 34 \\ \times \quad 7 \\ \hline \end{array}$$

3.
$$\begin{array}{r} 47 \\ \times \quad 8 \\ \hline \end{array}$$

4.
$$\begin{array}{r} 63 \\ \times \quad 9 \\ \hline \end{array}$$

5.
$$\begin{array}{r} 143 \\ \times \quad 5 \\ \hline \end{array}$$

6.
$$\begin{array}{r} 242 \\ \times \quad 6 \\ \hline \end{array}$$

7.
$$\begin{array}{r} 437 \\ \times \quad 7 \\ \hline \end{array}$$

8.
$$\begin{array}{r} 318 \\ \times \quad 9 \\ \hline \end{array}$$

9.
$$\begin{array}{r} 666 \\ \times \quad 7 \\ \hline \end{array}$$

10.
$$\begin{array}{r} 706 \\ \times \quad 9 \\ \hline \end{array}$$

UNIT 6

THE AREA METHOD

When multiplying two-digit numbers, the area method is a strategy that can simplify the multiplication by using a clear diagram of the process.

For example, to find **14 × 16** using the area method, draw a grid that is 14 by 16. Then **divide** the grid into **four parts** by **number splitting** 14 and 16 into 10 + 4 and 10 + 6.

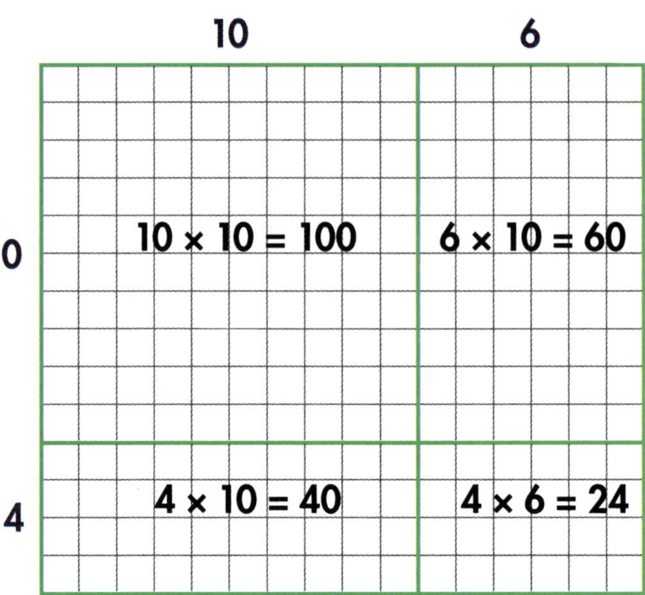

Can you see how the area method breaks the multiplication down into four easy parts?

The area method works by breaking down the multiplication into manageable parts.

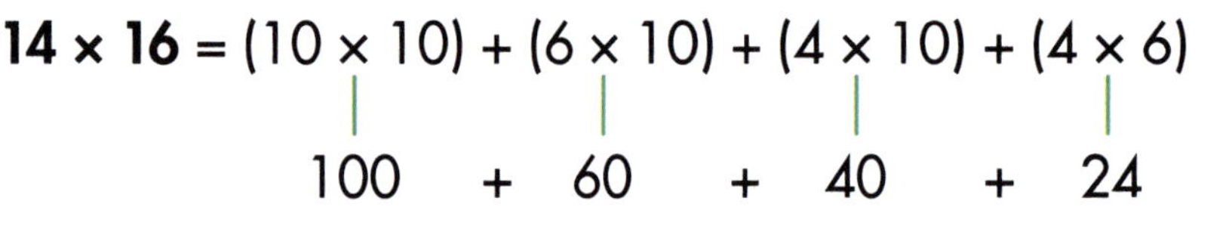

14 × 16 = (10 × 10) + (6 × 10) + (4 × 10) + (4 × 6)

100 + 60 + 40 + 24 = **224**

We practise

Write the multiplication and find the answer using the area method.

	20	6
10	20 x 10 = 200	6 x 10 = 60
2	20 x 2 = 40	6 x 2 = 12

The multiplication is 26 × 12

The parts are (20 × 10) + (6 × 10) + (20 × 2) + (6 × 2) = 312

You practise Write each multiplication and find the answer using the area method.

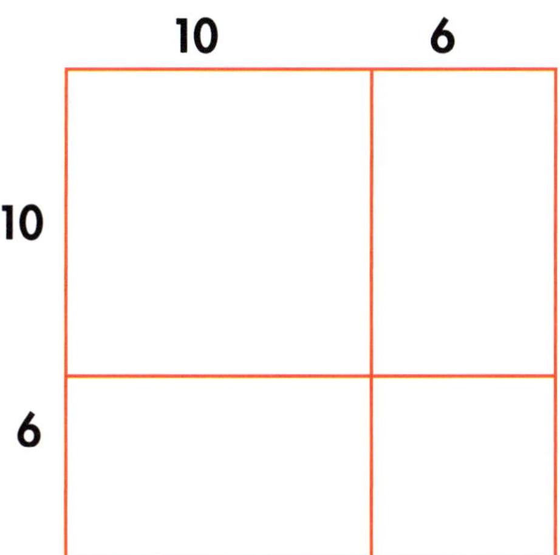

The multiplication is ___ × ___.

The parts are (___ × ___) + (___ × ___) + (___ × ___) + (___ × ___) = _____.

2

The multiplication is ___ × ___.

The parts are (___ × ___) + (___ × ___) + (___ × ___) + (___ × ___) = _____.

You practise On a separate piece of paper, draw the grids for these multiplications and then complete the number sentences.

17 × 16 = (___ × ___) + (___ × ___) + (___ × ___) + (___ × ___) = _____

23 × 18 = (___ × ___) + (___ × ___) + (___ × ___) + (___ × ___) = _____

13 × 24 = (___ × ___) + (___ × ___) + (___ × ___) + (___ × ___) = _____

33 × 14 = (___ × ___) + (___ × ___) + (___ × ___) + (___ × ___) = _____

15 × 31 = (___ × ___) + (___ × ___) + (___ × ___) + (___ × ___) = _____

82 × 11 = (___ × ___) + (___ × ___) + (___ × ___) + (___ × ___) = _____

BOB time!

THE GRID METHOD

When you use the area method, the grid diagram has to be to scale.

For example, for **23 × 13**, the diagram looks like this.

The grid method is similar to the area method. It also uses number splitting, but for the grid method, the diagram does not have to be to scale.

So for **23 × 13**, the diagram looks like this.

	20	3
10	200	30
3	60	9

I like to group some of the numbers in the grid first.

The **grid method** works for **larger numbers** too. For example, this grid shows how to find **124 × 13**.

	100	20	4
10	1000	200	40
3	300	60	12

Add the numbers in the grid for the answer.
1000 + 500 + 100 + 12 = 1612

We practise

Use the grid method to find 33 × 24.

	30	3
20	600	60
4	120	12

600 + 180 + 12 = 792

Use the grid method to find 254 × 23.

	200	50	4
20	4000	1000	80
3	600	150	12

4000 + 1600 + 230 + 12 = 5842

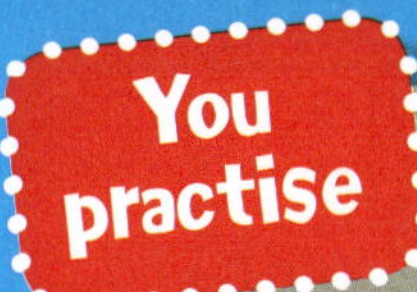

Complete these multiplications using the grid method.

24 × 16 = ______

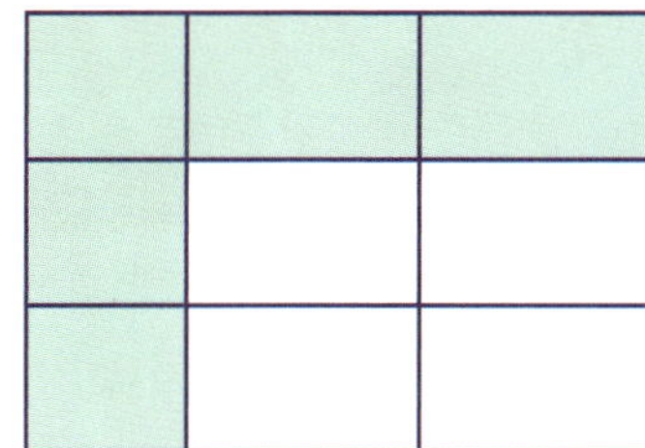

33 × 14 = ______

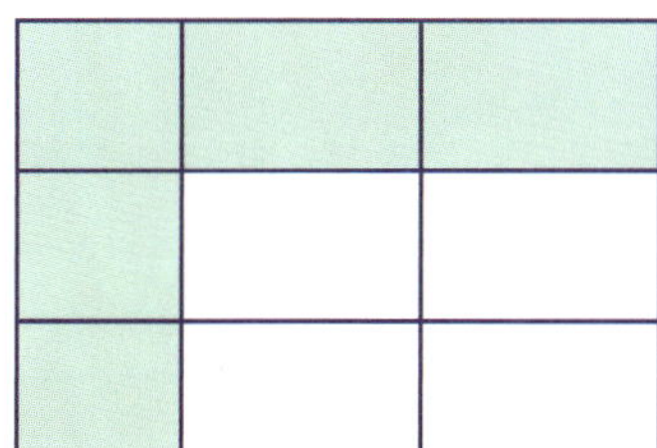

42 × 24 = ______

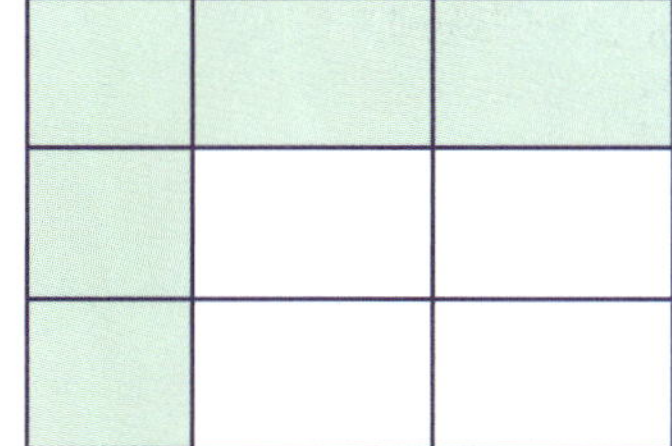

36 × 23 = ______

132 × 13 = ______

136 × 23 = ______

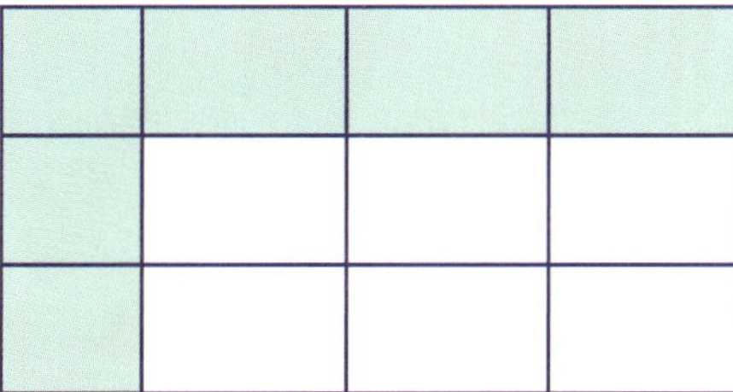

243 × 16 = ______

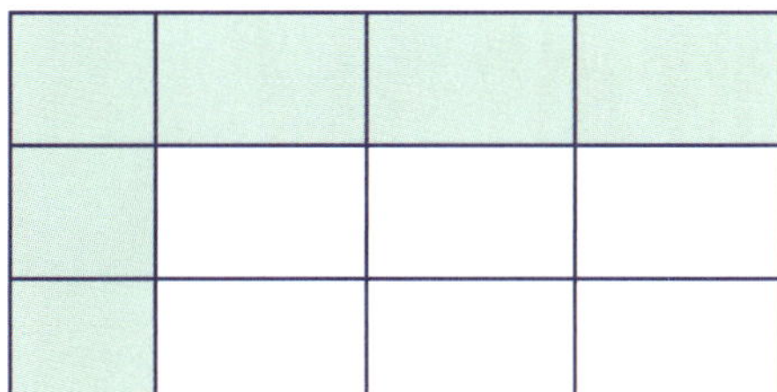

226 × 35 = ______

Remember to look for smart ways to add the numbers in the grid.

UNIT 8 EQUIVALENT DIVISIONS

When you are dividing a large number by eight, it is useful to think in terms of halving.

For example, to find **104 divided by 8,** you can use repeated halving:

104 ÷ 8 is equivalent to

52 ÷ 4, which is equivalent to

26 ÷ 2, which is equivalent to

13 ÷ 1 = 13

The answer is **13**.

If you are finding the halving difficult, use doubling and halving for multiplying and dividing by 2.

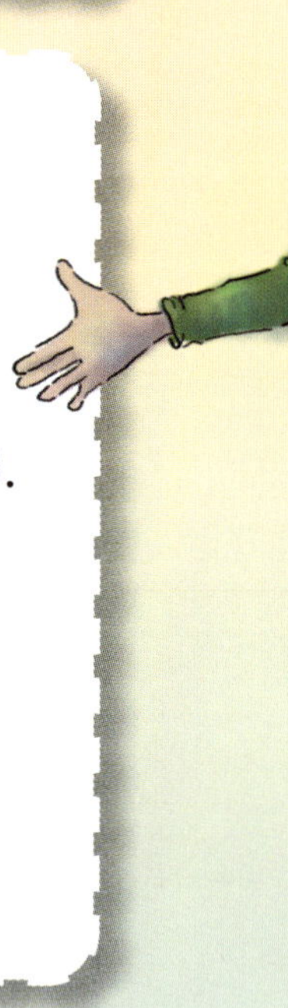

Whenever the divisor is a multiple of 2, then the repeated halving method often simplifies the division.

For example, to find **156 ÷ 12**, you can use **repeated halving** and finish with a **division by 3**.

156 ÷ 12 is equivalent to

78 ÷ 6, which is equivalent to

39 ÷ 3 = 13

The answer is **13**.

We practise

Use equivalent division to find 96 ÷ 8.

96 ÷ 8

48 ÷ 4

24 ÷ 2

12 ÷ 1 = 12

The answer is 12.

Use equivalent division to find 192 ÷ 12.

192 ÷ 12

96 ÷ 6

48 ÷ 3

16 ÷ 1 = 16

The answer is 16.

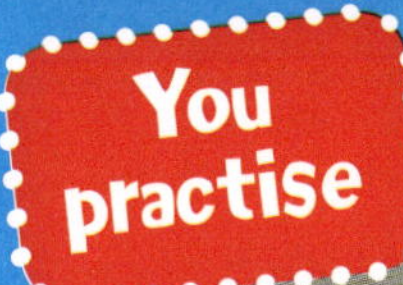

Use equivalent division for these questions.

52 ÷ 4

___ ÷ ___

___ ÷ ___

The answer is ____

96 ÷ 8

___ ÷ ___

___ ÷ ___

___ ÷ ___

The answer is ____

84 ÷ 6

___ ÷ ___

___ ÷ ___

The answer is ____

80 ÷ 16

___ ÷ ___

___ ÷ ___

___ ÷ ___

___ ÷ ___

The answer is ____

108 ÷ 12

___ ÷ ___

___ ÷ ___

___ ÷ ___

The answer is ____

136 ÷ 8

___ ÷ ___

___ ÷ ___

___ ÷ ___

The answer is ____

102 ÷ 6

___ ÷ ___

___ ÷ ___

The answer is ____

112 ÷ 16

___ ÷ ___

___ ÷ ___

___ ÷ ___

___ ÷ ___

The answer is ____

120 ÷ 8

___ ÷ ___

___ ÷ ___

___ ÷ ___

The answer is ____

228 ÷ 12

___ ÷ ___

___ ÷ ___

___ ÷ ___

The answer is ____

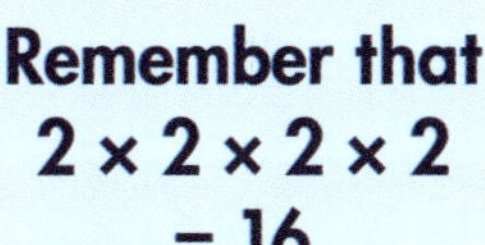

Remember that
2 × 2 × 2 × 2
= 16

Remember that
2 × 2 × 3
= 12

BOB time!

UNIT 9

DIVISION by NUMBER SPLITTING

To make the number split, think of multiples of the divisor – 4, 8, 12, 16, and so on.

Number splitting allows you to work with divisions that you already know.

For example, to work out **56 ÷ 4**, think about **4 × 10 = 40** and split **56** into **40 + 16.**

Look at this diagram to see how this works.

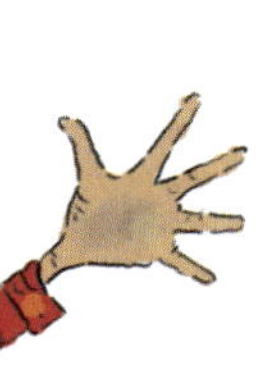

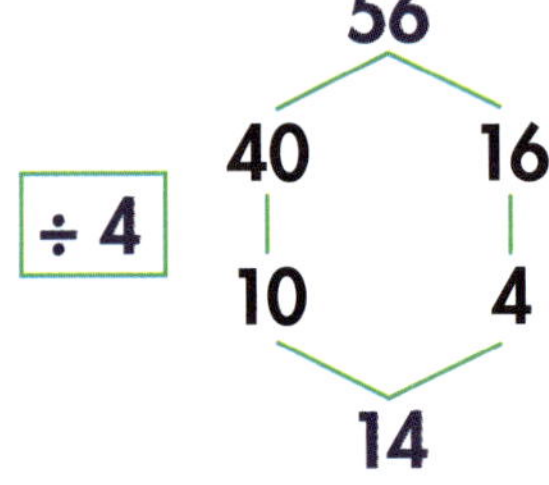

Number splitting works well for larger numbers too.

For example, you can use number splitting to work out 138 ÷ 6.

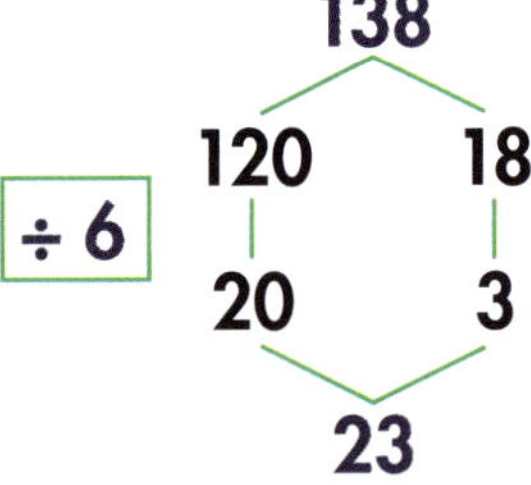

We practise

Use number splitting to work out these divisions.

72 ÷ 6

72

÷ 6

60 | 12

10 | 2

12

176 ÷ 8

176

÷ 8

160 | 16

20 | 2

22

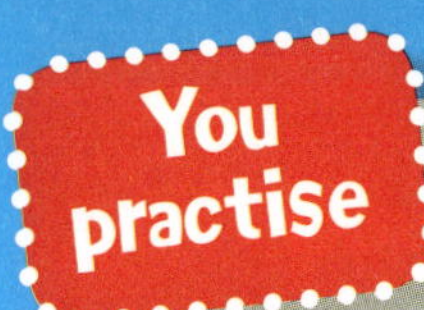

Use number splitting to work out these divisions.

42 ÷ 3

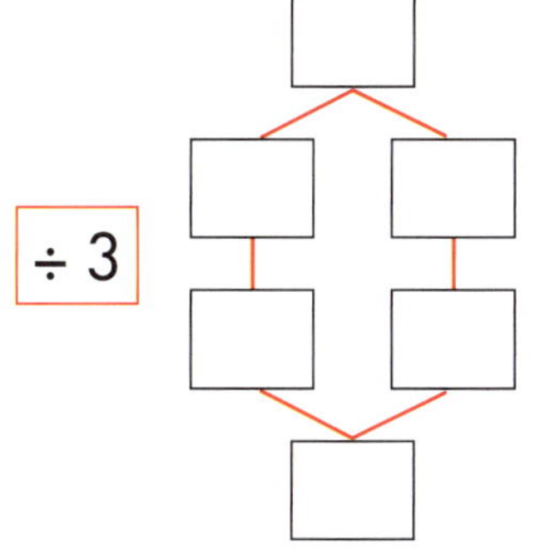

96 ÷ 8

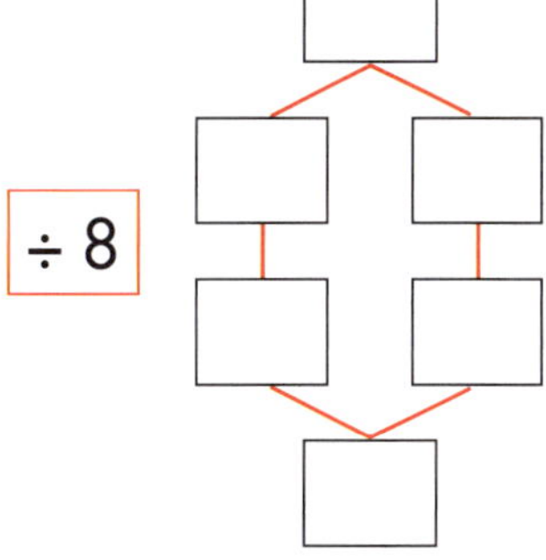

85 ÷ 5

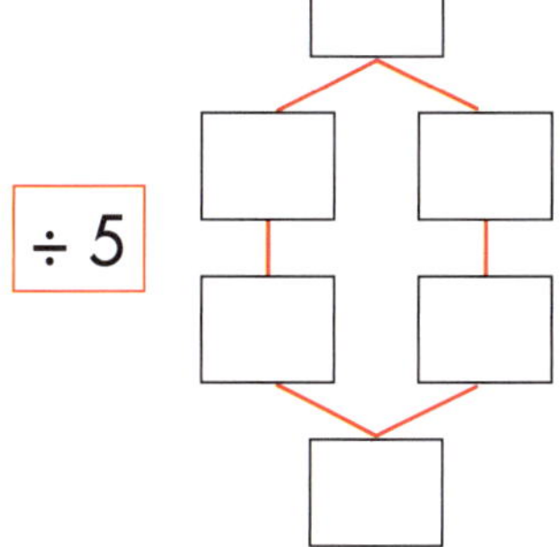

96 ÷ 4

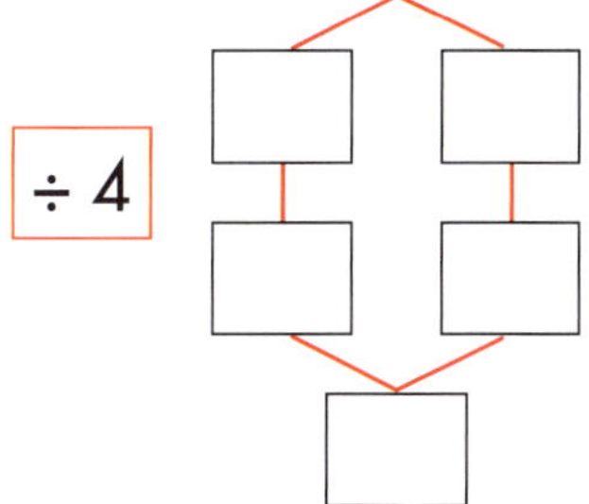

154 ÷ 7

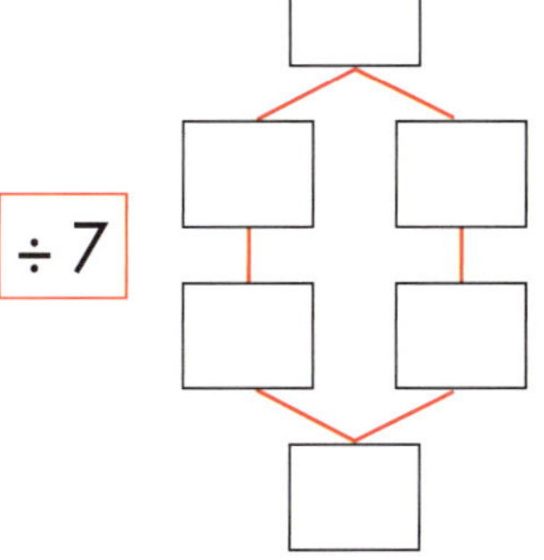

84 ÷ 7

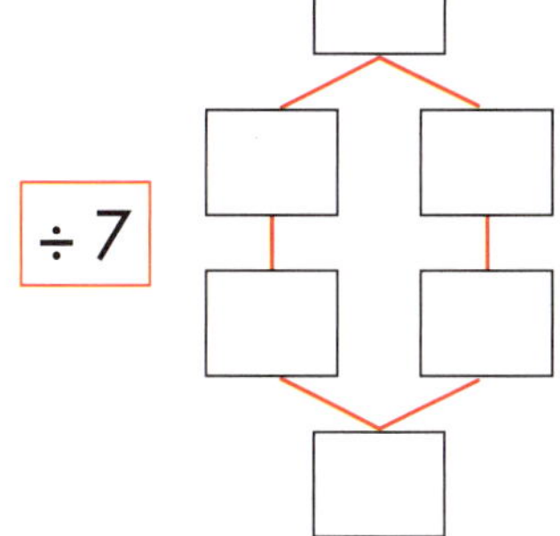

104 ÷ 8

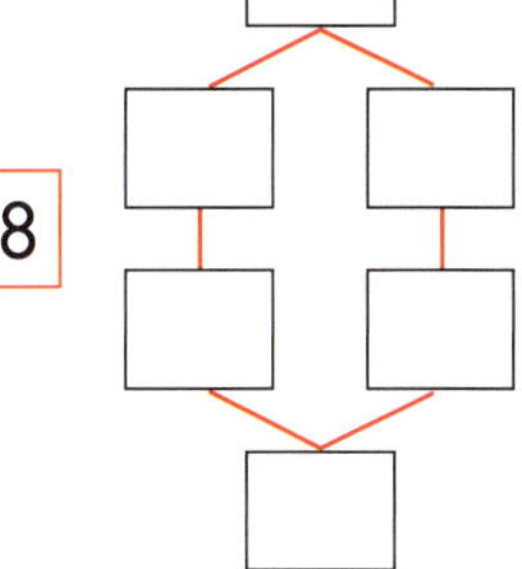

104 ÷ 4

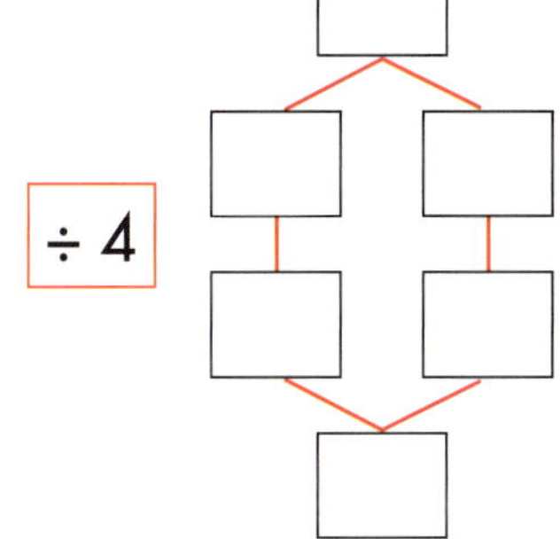

84 ÷ 6

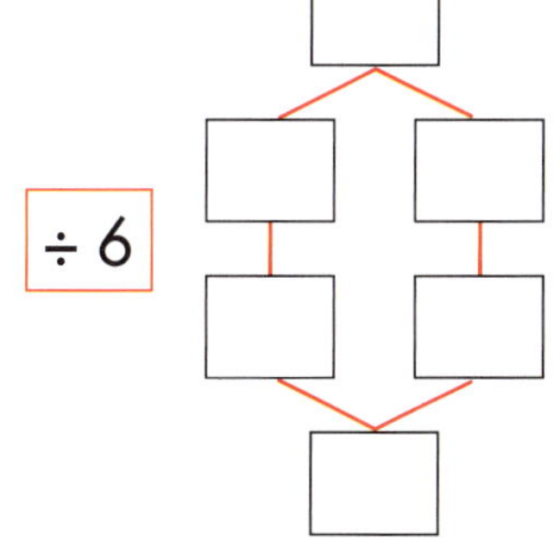

144 ÷ 6

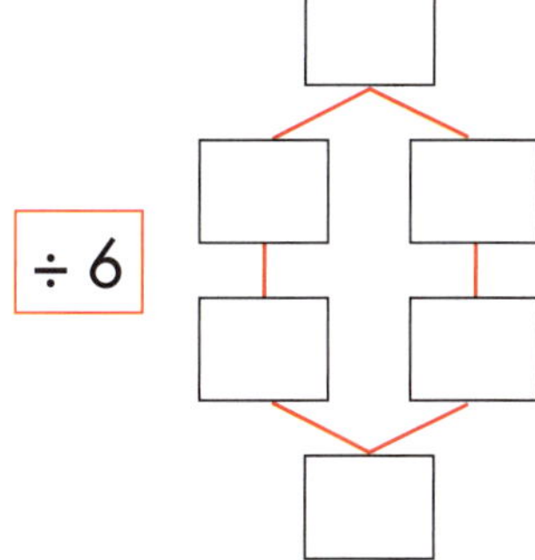

UNIT 10

PROBLEM SOLVING

How many muffins?

The baker was making muffins. She made sixteen trays, each with 12 muffins.

How many muffins did the baker make?

Notice that the important information is highlighted in blue and what has to be found out is highlighted in pink.

Use the information to create a visual picture and then it is easy to see what to do. The grid method and the area method are two ways that you could work out this problem.

Grid Method

	10	6
10	100	60
2	20	12

120 + 72 = 192

Area Method

	10	6	
10	100	60	160
2	20	12	32
			192

The baker made 192 muffins.

We practise

Highlight the important information and what you have to find out in this problem and then use the grid method to work out the answer.

Jake is a stamp collector and now has 22 pages with 14 stamps on each page.

How many stamps does Jake have altogether?

	20	2
10	200	20
4	80	8

Answer: Jake has 308 stamps.

Remember to give your answer as a sentence.

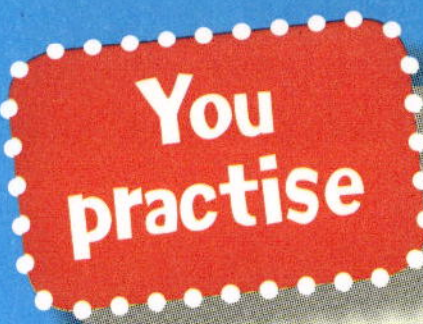

Highlight the important information and solve these problems using multiplication or division.

1. It takes 12 drinking straws to make a cube. How many straws do 26 children need to make one cube each?

2. If there are 147 days until Jake's birthday, how many weeks is that?

3. The plane has seats going across a row in 3 sets of 3. All 28 rows are full. How many passengers are on the plane?

4. The baker put 3, 4 or 5 sugar-flowers on each cupcake. If she makes 12 of each type of cupcake, how many sugar-flowers does she need?

5. Every month Clare saves $35. How much will she save by the end of the year if she starts saving at the beginning of February?

6. 'Share these 104 pencils equally between 8 art kits,' says Mrs Green. How many pencils will go into each kit?

7. 'I have eight pages of stamps, each with 28,' Clare said. 'I have 14 pages, each with 17 stamps,' Jake said. Who has the greater number of stamps?

8. My dog ate two biscuits per day and there are 13 biscuits in each packet. I want to buy enough for the month of April. How many packets will I have to buy?

BOB time!

RULES FOR DIVISIBILITY by 2, 4, 5 and 10

To find out if one number is evenly divisible by another number, **divisibility rules** can be used.

Divisible by 2

The **last digit** of the number must be **even**.

Divisible by 4

If the **10s digit is even**, the **last digit** must be a **0**, **4** or **8**.

If the **10s digit is odd**, the **last digit** must be **2** or **6**.

Divisible by 5

The **last digit** of the number must be either **5** or **0**.

Divisible by 10

The **last digit** of the number must be **0**.

To know if a number is divisible by 2, 5 or 10, you only need to **look at the last digit**.

Examples

Numbers	Divisible by 2?	Divisible by 4?	Divisible by 5?	Divisible by 10?
32	✓	✓	✗	✗
33	✗	✗	✗	✗
50	✓	✗	✓	✓
505	✗	✗	✓	✗
516	✓	✓	✗	✗

We practise

Which number is evenly divisible by 4?

123 ✗
234 ✗
456 ✓

Which number is evenly divisible by 2 and by 5?

120 ✓
345 ✗
456 ✗

Back to Basics

MULTIPLICATION & DIVISON

YEARS 5 and 6

Back to Basics
MULTIPLICATION & DIVISON
YEARS 5 and 6
Back to Basics
MULTIPLICATION & DIVISON
YEARS 5 and 6
Back to Basics
MULTIPLICATION & DIVISON
YEARS 5 and 6
Back to Basics
MULTIPLICATION & DIVISON
YEARS 5 and 6
Back to Basics
MULTIPLICATION & DIVISON
YEARS 5 and 6
Back to Basics
MULTIPLICATION & DIVISON
YEARS 5 and 6
Back to Basics
MULTIPLICATION & DIVISON
YEARS 5 and 6
Back to Basics
MULTIPLICATION & DIVISON
YEARS 5 and 6
Back to Basics
MULTIPLICATION & DIVISON
YEARS 5 and 6
Back to Basics
MULTIPLICATION & DIVISON
YEARS 5 and 6
Back to Basics
MULTIPLICATION & DIVISON
YEARS 5 and 6
Back to Basics
MULTIPLICATION & DIVISON
YEARS 5 and 6
Back to Basics
MULTIPLICATION & DIVISON
YEARS 5 and 6
Back to Basics
MULTIPLICATION & DIVISON
YEARS 5 and 6
Back to Basics
MULTIPLICATION & DIVISON
YEARS 5 and 6

22	33	44	55
26	39	52	65
34	51	68	85
38	57	76	95

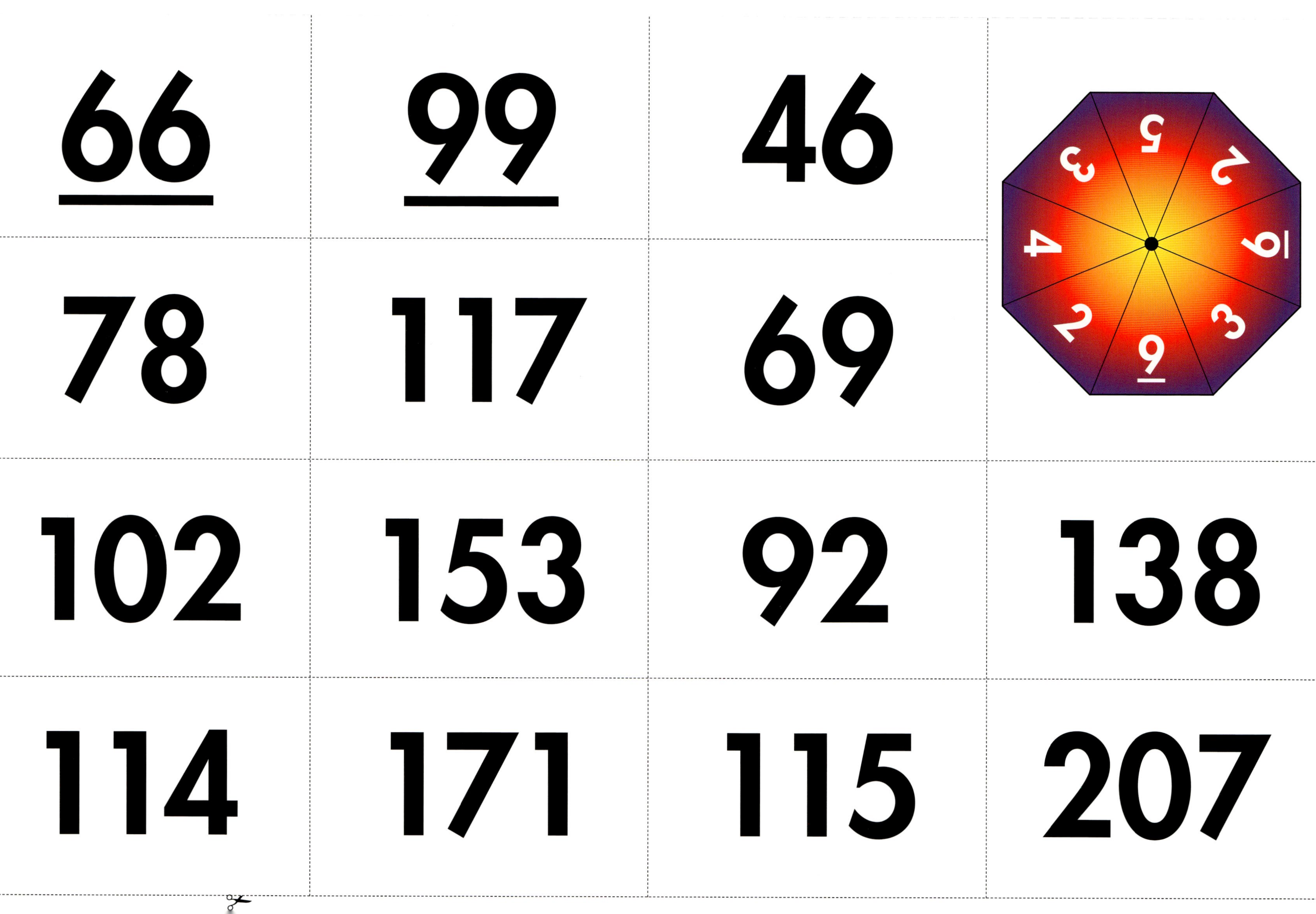
66
99
46
78
117
69
102
153
92
138
114
171
115
207

Back to Basics

MULTIPLICATION & DIVISON

YEARS 5 and 6

Back to Basics

MULTIPLICATION & DIVISON

YEARS 5 and 6

Back to Basics

MULTIPLICATION & DIVISON

YEARS 5 and 6

Back to Basics

MULTIPLICATION & DIVISON

YEARS 5 and 6

Back to Basics

MULTIPLICATION & DIVISON

YEARS 5 and 6

Back to Basics

MULTIPLICATION & DIVISON

YEARS 5 and 6

Back to Basics

MULTIPLICATION & DIVISON

YEARS 5 and 6

Back to Basics

MULTIPLICATION & DIVISON

YEARS 5 and 6

Back to Basics

MULTIPLICATION & DIVISON

YEARS 5 and 6

Back to Basics

MULTIPLICATION & DIVISON

YEARS 5 and 6

Back to Basics

MULTIPLICATION & DIVISON

YEARS 5 and 6

Back to Basics

MULTIPLICATION & DIVISON

YEARS 5 and 6

Back to Basics

MULTIPLICATION & DIVISON

YEARS 5 and 6

Back to Basics

MULTIPLICATION & DIVISON

YEARS 5 and 6

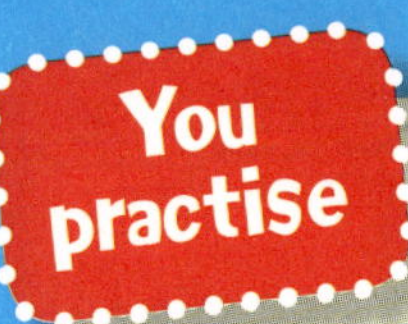

Circle Yes or No to answer these questions.

Is **368** evenly divisible by **4**? Yes No

Remember to use the divisibility rules.

Is **445** evenly divisible by **2**? Yes No

Is **464** evenly divisible by **4**? Yes No

4. Is **507** evenly divisible by **5**? Yes No

Is **6005** evenly divisible by **10**? Yes No

Is **326** evenly divisible by **2** and **4**? Yes No

Is **424** evenly divisible by **2** and **4**? Yes No

Is **1336** evenly divisible by **2** and **4**? Yes No

Is **370** evenly divisible by **2**, **5** and **10**? Yes No

Is **1025** evenly divisible by **2**, **5** and **10**? Yes No

BOB time!

DIVISIBILITY RULES for 3, 6 and 9

There are **three rules** that help you find out if a number is **evenly divisible by 3, 6 or 9**.

Divisible by 3
The **sum of the digits** of the number must be a **multiple of 3**.

Divisible by 6
The number must be **even** and the **sum of its digits** must be a **multiple of 3**.

Divisible by 9
The **sum of the digits** of the number must be a **multiple of 9**.

Examples

Numbers	Digit Sum	Divisible by 3?	Divisible by 6?	Divisible by 9?
32	5	✗	✗	✗
33	6	✓	✗	✗
54	9	✓	✓	✓
525	12 or 3	✓	✗	✗
9855	27 or 9	✓	✗	✓

If the digits add to a number that is more than 10, you can add those digits as well. For example, the digits of 9681 add to 24. Add again to get 6, which shows that **9681** is **divisible by 3**.

We practise

Are these numbers evenly divisible by 6?

124 ✗ No
345 ✗ No
456 ✓ Yes

Are these numbers evenly divisible by 9?

3456 ✓ Yes
5678 ✗ No
6786 ✓ Yes

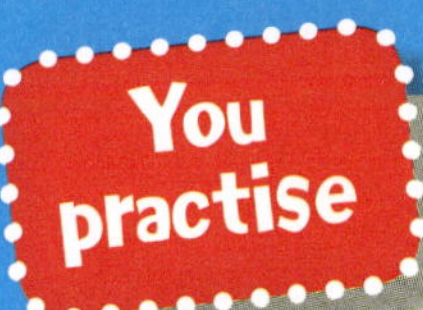

Circle Yes or No to answer these questions.

1. Is **368** evenly divisible by **6**? Yes No

2. Is **945** evenly divisible by **9**? Yes No

3. Is **464** evenly divisible by **3**? Yes No

4. Is **507** evenly divisible by **3** and **6**? Yes No

5. Is **6789** evenly divisible by **9**? Yes No

6. Is **2853** evenly divisible by **3** and **9**? Yes No

7. Is **2853** evenly divisible by **6** and **9**? Yes No

8. Is **24 858** evenly divisible by **3**, **6** and **9**? Yes No

9. Is **58 635** evenly divisible by **3**, **6** and **9**? Yes No

10. Is **123 876** evenly divisible by **3**, **6** and **9**? Yes No

Remember, first find the **sum of the digits**. You can check your answer on your calculator.

BOB time!

UNIT 13

FACTORS and MULTIPLES

Factors

2 × 3 = 6 1 × 6 = 6

1, 2, 3 and 6 divide into 6 without leaving a remainder. They are called factors of 6.

2 and 3 are also called **proper factors** of 6 because neither is equal to 1 or 6 itself.

You can use the **outside–inside method** to find all the **factors** of a given number.

For example, to find all the **factors of 24**, you list the factors in pairs.

To make the list, **start at 1** and link it to **24**, then add **2** to the list and link it to **12**.

If you reach a number that isn't a factor of 24, cross it out.

Stop when you reach 6, because it is already linked to a smaller number.

Multiples

2 × 3 = 6

2 × 4 = 8

2 × 5 = 10

6, 8 and 10 can all be made by multiplying a number by **2**. They are called **multiples of 2**.

We practise

Use the outside–inside method to find all the factors of 18.

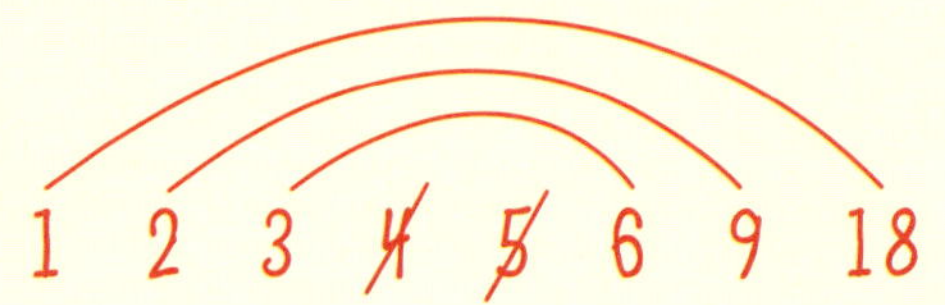

Which of these numbers are multiples of 3?

(3)	14	17	(21)	(33)
35	(39)	40	43	(48)

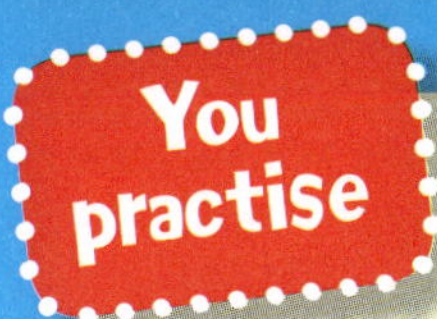

Use the outside–inside method to find all the factors of these numbers.

 15

 56

 28

 72

 48

 80

You practise

Circle the multiples.

 Which of these numbers are **multiples of 4**?

12 22 32 42 52 62 72 82 92

 Which of these numbers are **multiples of 6**?

16 26 36 46 56 66 76 86 96

 Which of these numbers are **multiples of 7**?

14 17 24 27 34 37 42 49 56 65

 Which of these numbers are **multiples of 11**?

31 44 87 99 133 147 154 188 209 222

BOB time!

UNIT 14

LONG DIVISION with REMAINDERS

Problem:

How many bags of 36 lollies can be made from this jar of 475 lollies?

To **solve the problem**, you need to **divide 475 by 36** using **long division**.
A good strategy is to think about taking bags out of the jar in a systematic way, using the multiples that you know.

		36) 4 7 5	
Step 1:	10 × 36	− 3 6 0	
		1 1 5	
Step 2:	2 × 36	− 7 2	
		4 3	
Step 3:	1 × 36	− 3 6	
Step 4:	**13 bags**	7	**lollies remaining**

Step 1: What is a large multiple of **36** that is easy to find and **less than 475**? **10 × 36 = 360**, so take 10 bags out of the jar.

Step 2: What is the next multiple of 36 that is easy to find and **less than 115**? **2 × 36 = 72**, so take another 2 bags out of the jar.

Step 3: **43** lollies in the jar, so you can take out 1 more bag.

Step 4: You can take **13 bags** altogether out of the jar and there will be **7 lollies left** in the jar. Write this as: **13 R 7**

We practise

Solve 296 ÷ 24 using long division.

	24) 2 9 6
10 × 24	− 2 4 0
	5 6
2 × 24	− 4 8
12	R 8

Solve 856 ÷ 35 using long division.

	35) 8 5 6
20 × 35	− 7 0 0
	1 5 6
2 × 35	− 7 0
	8 6
2 × 35	− 7 0
24	R 1 6

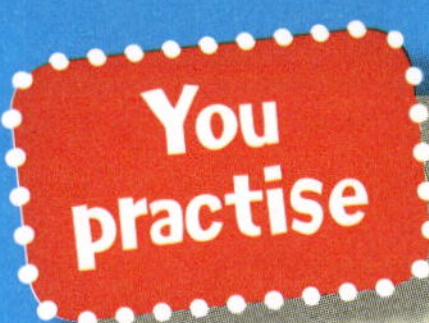

Complete these long divisions.

$25\overline{)305}$

$38\overline{)458}$

$47\overline{)568}$

$28\overline{)369}$

$37\overline{)250}$

$46\overline{)700}$

$18\overline{)434}$

$26\overline{)636}$

Remember – choose multiples that are easy to work out.

$33\overline{)961}$

$45\overline{)1950}$

BOB time!

ESTIMATE and CHECK

It is easy to press the wrong keys on your calculator, so always estimate the answer first.

It is always a good idea to estimate the answer to a multiplication before you work it out.

For example, to find **4 × 38**, follow these steps.

Step 1: Round the multiplicand, **38**, up to **40.**

Step 2: Work out **4 × 40 = 160** so you know the **answer** is close to, but less than, 160.

Step 3: Use your **calculator** to find **4 × 38 = 152**.

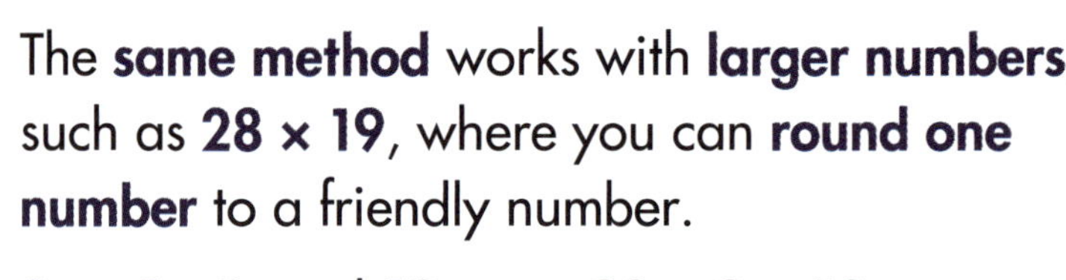

The **same method** works with **larger numbers**, such as **28 × 19**, where you can **round one number** to a friendly number.

Step 1: Round **19** up to **20 = 2 × 10**.

Step 2: **Double 28 = 56** and then **multiply by 10** to get **560**. You know the **answer** is close to, but less than, 560.

Step 3: Use your **calculator** to find **28 × 19 = 532**.

We practise

Estimate and check to work out 9 × 39.

Step 1: Round the multiplicand.

Round 39 to 40

Step 2: Make an estimate.

9 x 40 = 360

Step 3: Check on a calculator.

9 x 39 = 351

Estimate and check to work out 18 × 47.

Step 1: Round one of the numbers.

Round 18 to 20 = 2 × 10

Step 2: Make an estimate.

Double 47 = 94 and 94 x 10 = 940

Step 3: Check on a calculator.

18 x 47 = 846

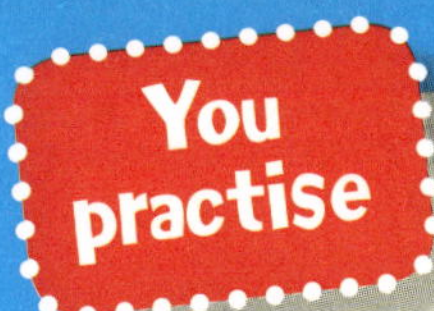

To find these multiplications, round the multiplicand, make an estimate and check on a calculator.

1 6×28

Step 1: ______________________

Step 2: ______________________

Step 3: ______________________

2 7×32

Step 1: ______________________

Step 2: ______________________

Step 3: ______________________

3 4×59

Step 1: ______________________

Step 2: ______________________

Step 3: ______________________

4 8×62

Step 1: ______________________

Step 2: ______________________

Step 3: ______________________

5 9×48

Step 1: ______________________

Step 2: ______________________

Step 3: ______________________

6 9×78

Step 1: ______________________

Step 2: ______________________

Step 3: ______________________

When you **round**, choose the number **closer** to a **friendly number**.

To find these multiplications, round one of the numbers, make an estimate and check on a calculator.

7 32×25

Step 1: ______________________

Step 2: ______________________

Step 3: ______________________

8 18×19

Step 1: ______________________

Step 2: ______________________

Step 3: ______________________

9 45×11

Step 1: ______________________

Step 2: ______________________

Step 3: ______________________

10 48×28

Step 1: ______________________

Step 2: ______________________

Step 3: ______________________

BOB time!

LANDMARK NUMBERS

The numbers **25**, **50**, **75** and **100** are called **landmark numbers** because they are important places on the road **from 0 to 100**.

At 75 you are $\frac{3}{4}$ of the way **to 100**.

The **landmark numbers** are related to each other by **multiplication and division**.

4 × 25 = 100	4 × 250 = 1000
100 ÷ 4 = 25	1000 ÷ 4 = 250
3 × 25 = 75	3 × 250 = 750
75 ÷ 3 = 25	750 ÷ 3 = 250
5 × 25 = 125	6 × 25 = 150
125 ÷ 5 = 25	150 ÷ 6 = 25

Landmark numbers are very useful when you need to make a **mental calculation**.

For example, to work out **4 × 27**, split 27 into the **landmark number 25 + 2**.

4 × 27 = 4 × 25 + 4 × 2
= 100 + 8
= **108**

We practise

Use a landmark number to work out 2 × 78.

2 × 78 = 2 × 75 + 2 × 3
= 150 + 6
= 156

Use a landmark number to work out 156 ÷ 6.

156 ÷ 6 = 150 ÷ 6 + 6 ÷ 6
= 25 + 1
= 26

You practise

Use a landmark number to work out each multiplication.

1. $4 \times 76 =$ ______________________
   $=$ ______________________
   $=$ ______________________

2. $4 \times 54 =$ ______________________
   $=$ ______________________
   $=$ ______________________

3. $3 \times 27 =$ ______________________
   $=$ ______________________
   $=$ ______________________

4. $6 \times 52 =$ ______________________
   $=$ ______________________
   $=$ ______________________

5. $5 \times 73 =$ ______________________
   $=$ ______________________
   $=$ ______________________

6. $6 \times 23 =$ ______________________
   $=$ ______________________
   $=$ ______________________

You practise

Use a landmark number to work out each division.

7. $204 \div 4 =$ ______________________
   $=$ ______________________
   $=$ ______________________

8. $159 \div 3 =$ ______________________
   $=$ ______________________
   $=$ ______________________

9. $130 \div 5 =$ ______________________
   $=$ ______________________
   $=$ ______________________

10. $162 \div 6 =$ ______________________
    $=$ ______________________
    $=$ ______________________

BOB time!

IS IT ALWAYS TRUE?

A **mathematical statement** can be:

- **Always true**
- **Sometimes true**
- **Never true**

When you multiply 12 by 3 the answer is 36. 36 is larger than 12, so multiplying by 3 made 12 larger.

With **multiplication** and **division**, there are two statements that people think are **always true**:

1. When you **multiply** a number, you make it **larger**.
2. When you **divide** a number you make it **smaller**.

But wait! When you **multiply 12 by 0·5** the answer is **6**, which is **half** of 12.

12 × 0·5 = 6

So if you think about this multiplication as '12 halves', as in the diagram below, then you will get the answer **6**, which is **smaller than 12**.

This diagram also shows 6 circles that have been **divided into halves**.
This means that when you **divide 6 by 0·5**, the answer is **12**, which is larger than **6**.

6 ÷ 0·5 = 12

We practise

Complete this multiplication and decide whether the statement is always, sometimes or never true.

12 × 4 = 48

When you multiply a number by 4 it gets larger.

Sometimes Never

Complete this division and decide whether the statement is always, sometimes or never true.

18 ÷ 0·5 = 36

When you divide a number by 0·5 it gets smaller.

Always Sometimes

You practise

Complete each multiplication and division and decide whether the statement is always, sometimes or never true.

15 × 3 = ______

When you multiply a number by 3 it gets larger.

Always Sometimes Never

18 ÷ 3 = ______

When you divide a number by 3 it gets smaller.

Always Sometimes Never

24 × 0·5 = ______

When you multiply a number by 0·5 it gets larger.

Always Sometimes Never

7 ÷ 0·5 = ______

When you divide a number by 0·5 it gets smaller.

Always Sometimes Never

16 × 0·25 = ______

When you multiply a number by 0·25 it gets smaller.

Always Sometimes Never

12 ÷ 0·5 = ______

When you divide a number by 0·5 it gets larger.

Always Sometimes Never

30 ÷ 0·3 = ______

When you divide a number by 0·3 it gets larger.

Always Sometimes Never

7 ÷ 0·1 = ______

When you divide a number by 0·1 it gets larger.

Always Sometimes Never

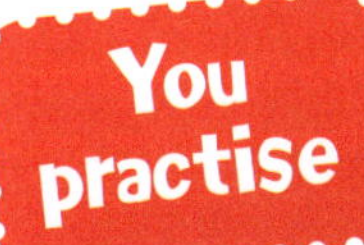

Are these statements always, sometimes or never true?

When you multiply a number, you make it larger.

Always Sometimes Never

When you divide a number you make it smaller.

Always Sometimes Never

BOB time!

SIGNALS TO DIVIDE

There are several different ways that divisions can be expressed.

These different **expressions** all mean the same thing; they are all **signals to divide**.

$36 \div 6 \qquad \frac{36}{6} \qquad 36 \times \frac{1}{6} \qquad 6\overline{)36}$

$\frac{36}{6}$ looks like a fraction, doesn't it? But it is telling you to divide 36 by 6.

Dividing by 6 and multiplying by the fraction $\frac{1}{6}$ are the same thing, but it's easiest to think of them as **divide by 6**.

Sometimes you are asked to find the value of a fraction such as:

$$\frac{34 + 15}{7}$$

Rather than write this as two separate fractions like $\frac{34}{7} + \frac{15}{7}$, it is easiest to add the numerators first and then do the division.

$$\frac{34 + 15}{7} = \frac{49}{7} = 7$$

We practise

Write these as 'divide by' expressions and work out their values.

$\frac{24}{8} = 24 \div 8 = 3$

$28 \times \frac{1}{7} = 28 \div 7 = 4$

$7\overline{)35} = 35 \div 7 = 5$

What is the value of this fraction?

$\frac{26 + 19}{9} = \frac{45}{9} = 5$

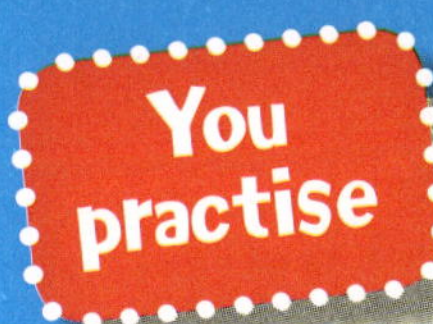

Write each division as a 'divide by' expression and work out its value.

 $\frac{39}{3}$

= __________ = ____

 $54 \times \frac{1}{9}$

= __________ = ____

 $\frac{72}{8}$

= __________ = ____

 $9\overline{)108}$

= __________ = ____

 $42 \times \frac{1}{6}$

= __________ = ____

 $24\overline{)336}$

= __________ = ____

You might want to look back to Unit 14 where you learnt about long division.

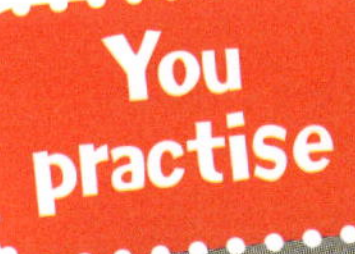

What is the value of each fraction?

 $\frac{15 + 27}{7}$

= _____ = _____

 $\frac{51 + 57}{12}$

= _____ = _____

 $\frac{39 + 42}{9}$

= _____ = _____

10 $\frac{178 + 47}{15}$

= _____ = _____

BOB time!

MULTIPLY and DIVIDE DECIMALS by POWERS OF 10

In Unit 2 you learnt about multiplying whole numbers by 10 or 100.

The same process works when you **multiply a decimal number by 10 or 100.**

For example, look at **37·42 × 10** on a place-value diagram.

100	10	1	$\frac{1}{10}$	$\frac{1}{100}$	
	3	7 ·	4	2	**37·42 × 10**
3	7	4 ·	2		**= 374·2**

Each digit of 37·42 moves **one place-value position to the left** when you multiply by 10.

I would check these on my calculator.

Dividing by 10 or 100 is just the reverse. For example, look at **415·6 ÷ 100** on a place-value diagram.

100	10	1	$\frac{1}{10}$	$\frac{1}{100}$	$\frac{1}{1000}$	
4	1	5 ·	6			**415·6 ÷ 100**
		4 ·	1	5	6	**= 4·156**

Each digit of 415·6 moves **two place-value positions to the right** when you divide by 100.

We practise

Use a place-value diagram to find 42·76 × 10.

100	10	1	$\frac{1}{10}$	$\frac{1}{100}$	
	4	2 ·	7	6	42·76 × 10
4	2	7 ·	6		= 427·6

Use a place-value diagram to find 572·3 ÷ 10.

100	10	1	$\frac{1}{10}$	$\frac{1}{100}$	
5	7	2 ·	3		572·3 ÷ 10
	5	7 ·	2	3	= 57·23

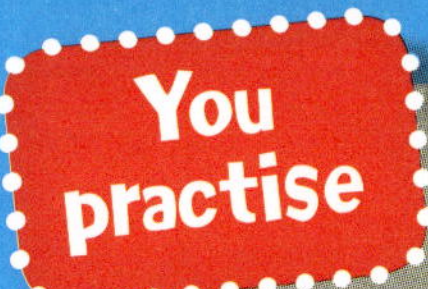

Use a place-value diagram to find these multiplications.

4·56 × 10 = ________

100	10	1	$\frac{1}{10}$	$\frac{1}{100}$

5·27 × 100 = ________

100	10	1	$\frac{1}{10}$	$\frac{1}{100}$

27·2 × 10 = ________

100	10	1	$\frac{1}{10}$	$\frac{1}{100}$

6·09 × 100 = ________

100	10	1	$\frac{1}{10}$	$\frac{1}{100}$

You practise

Use a place-value diagram to find these divisions.

47·6 ÷ 10 = ________

100	10	1	$\frac{1}{10}$	$\frac{1}{100}$	$\frac{1}{1000}$

305·7 ÷ 100 = ________

100	10	1	$\frac{1}{10}$	$\frac{1}{100}$	$\frac{1}{1000}$

915·3 ÷ 10 = ________

100	10	1	$\frac{1}{10}$	$\frac{1}{100}$	$\frac{1}{1000}$

970 ÷ 1000 = ________

100	10	1	$\frac{1}{10}$	$\frac{1}{100}$	$\frac{1}{1000}$

BOB time!

How many numbers?

Mrs Green asked her class, 'How many numbers between 80 and 100 are multiples of 3, 4 or 5?'

How many numbers do you think they found?

Notice that the important information is highlighted in blue and what has to be found out is highlighted in pink.

You can use the information to make a visual picture as that can often help you solve the problem. For this problem, you could draw a table and tick each of the numbers that is a multiple of 3.

	81	82	83	84	85	86	87	88	89	90	91	92	93	94	95	96	97	98	99
3	✓			✓			✓			✓			✓			✓			✓

Complete the solution by placing a tick against each number that is a multiple of 4 and each number that is a multiple of 5.

	81	82	83	84	85	86	87	88	89	90	91	92	93	94	95	96	97	98	99
3	✓			✓			✓			✓			✓			✓			✓
4				✓				✓				✓				✓			
5					✓					✓					✓				

How many numbers did Mrs Green's class find? 14

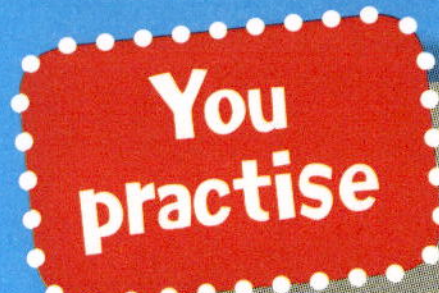

Highlight the important information and solve these problems.

1. Jake thought of a number. He said that his number is a multiple of 4, and 1 more than a multiple of 5. What is his number?

___

2. Clare said that her number is a multiple of 2, 4 and 5, between 30 and 70 and as large as possible. What is her number?

___

3. 'I'm thinking of a number between 120 and 140,' said Clare. 'It is a multiple of 9, but not a multiple of 5.' What is Clare's number?

___

4. Clare asked Jake if 1234 could be divided by 4. He said, 'Yes, it can.' What do you think?

___

5. Which one of these numbers is divisible by both 4 and 9?

1344 1233 3132 4122

6. Max arranged his stamp collection to have 16 stamps on each page. He has 184 stamps in his collection. How many pages does he have?

___

7. The dog breeder bought 19 bags of dog biscuits at $27 each. She counted the $50 notes in her purse and found that she had 11 of them.

Will she be able to afford the dog biscuits and, if so, how much change should she expect?

___

8. 252 helium balloons need to be made into bunches of 12 for the display.

How many bunches can be made?

___

9. When you divide a number by 0.25, it gets smaller.

Always Sometimes Never

10. Clare said that she could write 'forty eight shared between eight' in four different ways.

How can she do that?

___

___

___

___

BOB time!

TEST 1

Use number splitting to halve 76.

64 × 100 = ________

300 × 120 = ________

Split the multiplicand to find 4 × 27.

Split the multiplier to find 6 × 24.

Use chunking to find 58 × 6.

Write the multiplication and find the answer using this area diagram.

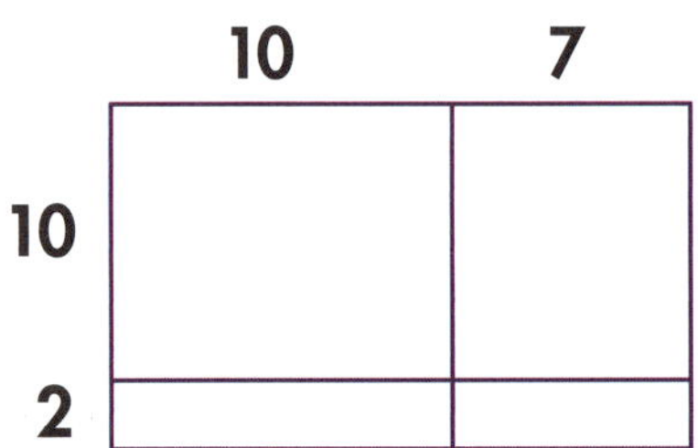

____ × ____ = ______

Use the multiplication grid to find

24 × 16 = ________

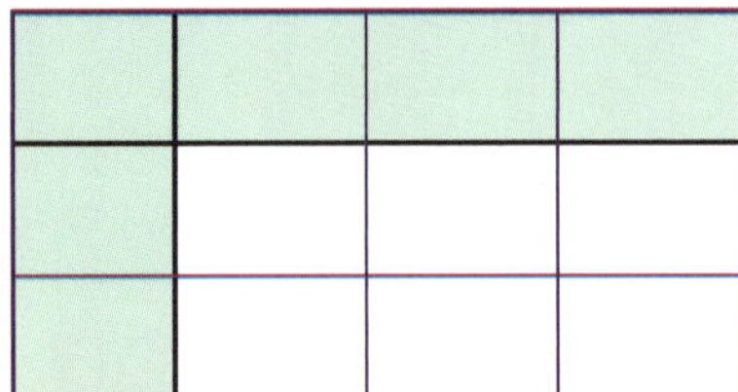

Use the front-end method to find 417 × 6.

Complete these equivalent divisions.

52 ÷ 4

____ ÷ ____

____ ÷ ____

The answer is ____

Use number splitting to find 96 ÷ 8.

Highlight the important information in this problem and use a multiplication grid to find the answer.

A gardener planted 17 rows of potatoes and found that he could fit 14 into each row.

How many potatoes did the gardener plant? ________

TEST 2

Circle the numbers that are multiples of 4.

28, 32, 38, 42, 48, 56, 66, 84

Which of these numbers are evenly divisible by 4 and 5?

30, 50, 60, 370, 580, 690, 1000

Circle any of these numbers that are evenly divisible by 3 and 6.

1233, 2344, 3456, 4563

Complete this long division.

$32 \overline{)702}$

To find 21 × 58, round one of the numbers, make an estimate and check on your calculator.

Step 1: ______________________

Step 2: ______________________

Step 3: ______________________

____

Use a landmark number to find 6 × 24.

6 × 24 = ______________________

= ______________________

= ______________________

Use a landmark number to find 182 ÷ 7.

182 ÷ 7 = ______________________

= ______________________

= ______________________

When you multiply a number, you make it larger. Is this:

- ◯ always true?
- ◯ sometimes true?
- ◯ never true?

$72 \times \frac{1}{8} =$ ______ ÷ ______ = ______

$\frac{13}{7} + \frac{22}{7} =$ ______ ÷ ______ = ______

34·07 × 100 = __________

205·6 ÷ 100 = __________

Highlight the important information in this problem and use a table to find the answer.

How many numbers between 150 and 170 are multiples of 3, 4 or 5?

____

ANSWERS

Unit 1

1

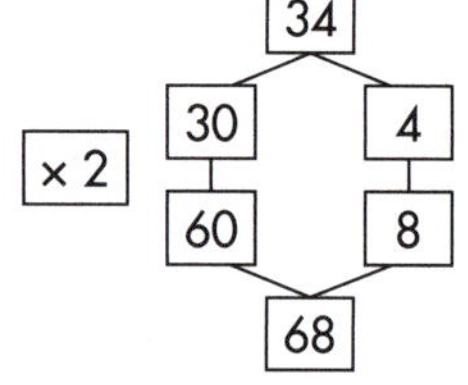

2

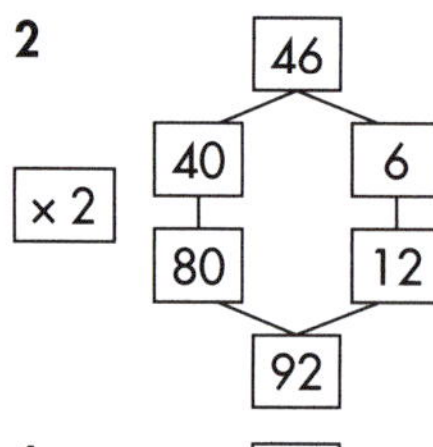

3

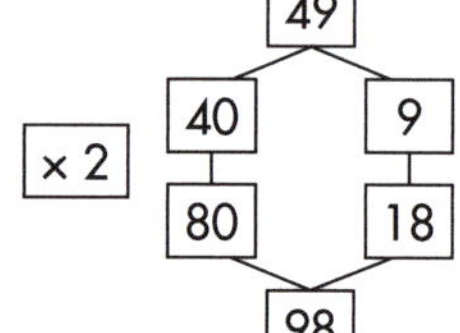

4

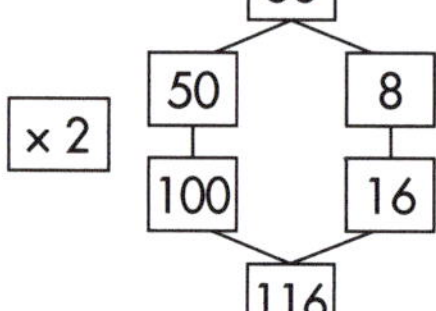

5

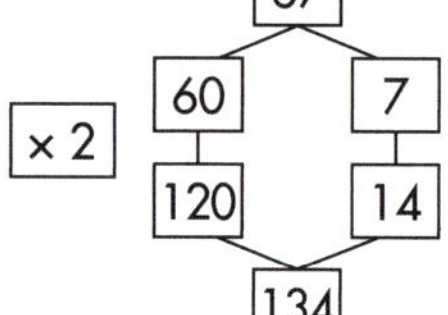

6

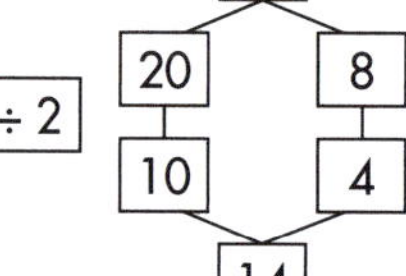

7

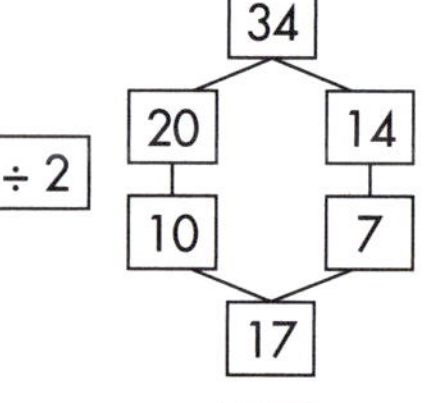

8

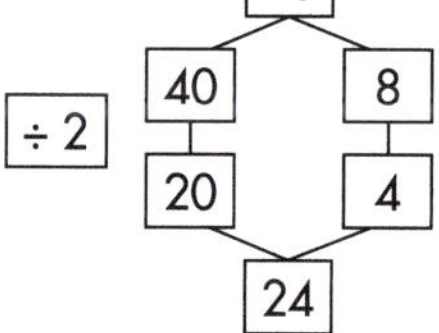

9

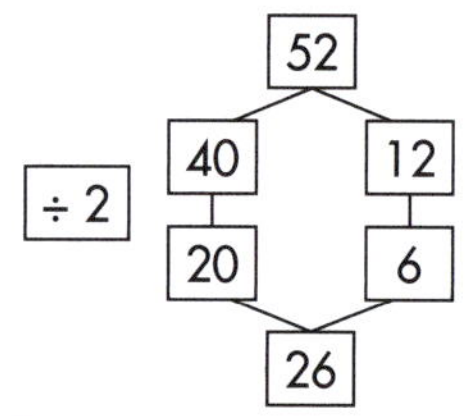

10

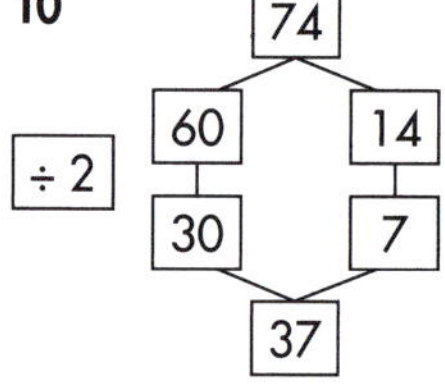

Unit 2

1 50
2 350
3 1080
4 560
5 8500
6 1500
7 5600
8 2800
9 180 000
10 280 000

Unit 3

1 3 × 15 ⟨ 3 × 10 = 30 ; 3 × 5 = 15 ⟩ 45

2 4 × 32 ⟨ 4 × 30 = 120 ; 4 × 2 = 8 ⟩ 128

3 5 × 43 ⟨ 5 × 40 = 200 ; 5 × 3 = 15 ⟩ 215

4 6 × 52 ⟨ 6 × 50 = 300 ; 6 × 2 = 12 ⟩ 312

5 7 × 23 ⟨ 7 × 20 = 140 ; 7 × 3 = 21 ⟩ 161

6 6 × 14 = 2 × 3 × 14
42
84

7 4 × 29 = 2 × 2 × 29
58
116

8 6 × 43 = 2 × 3 × 43
129
258

9 8 × 15 = 2 × 4 × 15
60
120

10 8 × 73 = 2 × 2 × 2 × 73
146
292
584

Unit 4

1 23 × 9
180 27
207

2

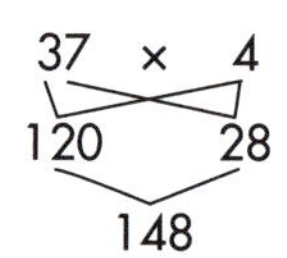

3 36 × 6
180 36
216

4 53 × 7
350 21
371

5 72 × 6
420 12
432

6 83 × 9
720 27
747

7 73 × 6
420 18
438

8

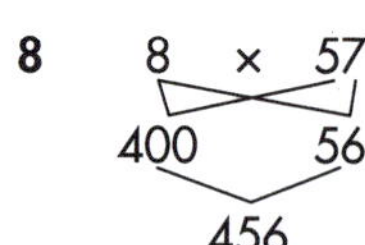

9 6 × 210
1200 60
1260

10 9 × 470
3600 630
4230

Unit 5

1

	2	3
×		4
	8	0
	1	2
	9	2

2

		3	4
×			7
	2	1	0
		2	8
	2	3	8

ANSWERS

3

	4	7
×		8
3	2	0
	5	6
3	7	6

4

	6	3
×		9
5	4	0
	2	7
5	6	7

5

		1	4	3
×				5
		5	0	0
		2	0	0
			1	5
		7	1	5

6

		2	4	2
×				6
	1	2	0	0
		2	4	0
			1	2
	1	4	5	2

7

	4	3	7
×			7
2	8	0	0
	2	1	0
		4	9
3	0	5	9

8

	3	1	8
×			9
2	7	0	0
		9	0
		7	2
2	8	6	2

9

	6	6	6
×			7
4	2	0	0
	4	2	0
		4	2
4	6	6	2

10

	7	0	6
×			9
6	3	0	0
		5	4
6	3	5	4

Unit 6

1

	10	6	
10	10 × 10 = 100	6 × 10 = 60	160
6	10 × 6 = 60	6 × 6 = 36	96
			256

The multiplication is 16 × 16.

The parts are (10 × 10) + (6 × 10) + (10 × 6) + (6 × 6) = 256.

2

	20	2	
10	20 × 10 = 200	2 × 10 = 20	220
2	20 × 2 = 40	2 × 2 = 4	44
			264

The multiplication is 22 × 12.

The parts are (20 × 10) + (2 × 10) + (20 × 2) + (2 × 2) = 264.

3 17 × 16 = (10 × 10) + (7 × 10) + (6 × 10) + (6 × 7) = 272

4 23 × 18 = (20 × 10) + (3 × 10) + (8 × 20) + (8 × 3) = 414

5 13 × 24 = (10 × 20) + (3 × 20) + (4 × 10) + (4 × 3) = 312

6 33 × 14 = (30 × 10) + (3 × 10) + (4 × 30) + (4 × 3) = 462

7 15 × 31 = (10 × 30) + (5 × 30) + (1 × 10) + (1 × 5) = 465

8 82 × 11 = (80 × 10) + (2 × 10) + (1 × 80) + (1 × 2) = 902

Unit 7

1 384
2 462
3 1008
4 828
5 1716
6 3128
7 3888
8 7910

Unit 8

1 13
2 12
3 14
4 5
5 9
6 17
7 17
8 7
9 15
10 19

Unit 9

1 14
2 17
3 12
4 14
5 12
6 24
7 13
8 24

ANSWERS

9 22
10 26

Unit 10

1 312
2 21
3 252
4 144
5 $385
6 13
7 Jake
8 5

Unit 11

1 Yes
2 No
3 Yes
4 No
5 No
6 No
7 Yes
8 Yes
9 Yes
10 No

Unit 12

1 No
2 Yes
3 No
4 No
5 No
6 Yes
7 No
8 Yes
9 No
10 Yes

Unit 13

1
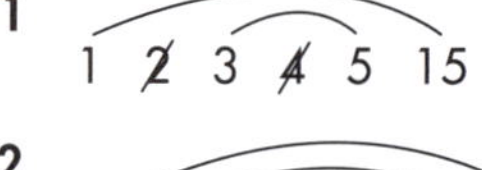

2
1 2 ~~3~~ 4 ~~5~~ ~~6~~ 7 14 28

3

4
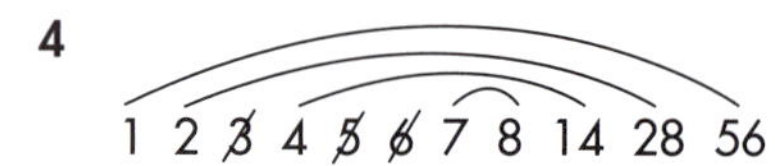

5
1 2 3 4 ~~5~~ 6 ~~7~~ 8 9 12 18 24 36 72

6
1 2 ~~3~~ 4 5 ~~6~~ ~~7~~ 8 ~~9~~ 10 16 20 40 80

7 12, 32, 52, 72, 92
8 36, 66, 96
9 14, 42, 49, 56
10 44, 99, 154, 209

Unit 14

1 12 R 5
2 12 R 2
3 12 R 4
4 13 R 5
5 6 R 28
6 15 R 10
7 24 R 2
8 24 R 12
9 29 R 4
10 43 R 15

Unit 15

1 Round 28 to 30, estimate 6 × 30 = 180, actual 168
2 Round 32 to 30, estimate 7 × 30 = 210, actual 224
3 Round 59 to 60, estimate 4 × 60 = 240, actual 236
4 Round 62 to 60, estimate 8 × 60 = 480, actual 496
5 Round 48 to 50, estimate 9 × 50 = 450, actual 432
6 Round 78 to 80, estimate 9 × 80 = 720, actual 702
7 Round 32 to 30, estimate 30 × 25 = 750, actual 800
8 Round 19 to 20, estimate 18 × 20 = 360, actual 342
9 Round 11 to 10, estimate 45 × 10 = 450, actual 495
10 Round 48 to 50, estimate 50 × 28 = 1400, actual 1344

Unit 16

1 4 × 76 = 4 × 75 + 4 × 1 = 300 + 4 = 304
2 4 × 54 = 4 × 50 + 4 × 4 = 200 + 16 = 216
3 3 × 27 = 3 × 25 + 3 × 2 = 75 + 6 = 81
4 6 × 52 = 6 × 50 + 6 × 2 = 300 + 12 = 312
5 5 × 73 = 5 × 75 − 5 × 2 = 375 − 10 = 365
6 6 × 23 = 6 × 25 − 6 × 2 = 150 − 12 = 138
7 204 ÷ 4 = 200 ÷ 4 + 4 ÷ 4 = 50 + 1 = 51
8 159 ÷ 3 = 150 ÷ 3 + 9 ÷ 3 = 50 + 3 = 53
9 130 ÷ 5 = 125 ÷ 5 + 5 ÷ 5 = 25 + 1 = 26
10 162 ÷ 6 = 150 ÷ 6 + 12 ÷ 6 = 25 + 2 = 27

ANSWERS

Unit 17

1. 45 Always
2. 6 Always
3. 12 Never
4. 14 Never
5. 4 Always
6. 24 Always
7. 100 Always
8. 70 Always
9. Sometimes
10. Sometimes

Unit 18

1. $39 \div 3 = 13$
2. $72 \div 8 = 9$
3. $42 \div 6 = 7$
4. $54 \div 9 = 6$
5. $108 \div 9 = 12$
6. $336 \div 24 = 14$
7. $42 \div 7 = 6$
8. $81 \div 9 = 9$
9. $108 \div 12 = 9$
10. $225 \div 15 = 15$

Unit 19

1. 45.6
2. 272
3. 527
4. 609
5. 4·76
6. 91·53
7. 3·057
8. 0·97

Unit 20

1. 36
2. 60
3. 126
4. No, it can't.
5. 3132
6. 12
7. Yes: $37
8. 21
9. Never
10. $\frac{48}{8}$, $48 \div 8$, $48 \times \frac{1}{8}$, $8\overline{)48}$

Test 1

1\.

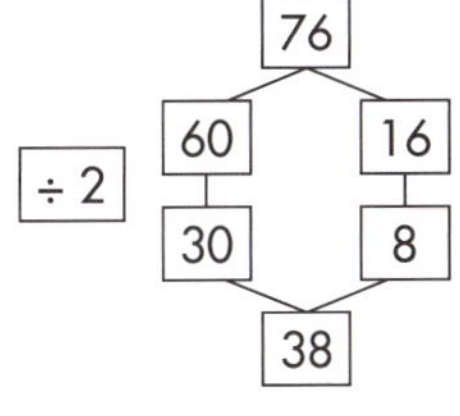

2\. 6400, 36 000

3\. $4 \times 27 = 4 \times 25 + 4 \times 2 = 100 + 8 = 108$
$6 \times 24 = 2 \times 3 \times 24 = 2 \times 72 = 144$

4\. 58 × 6
300 48
348

5\. $17 \times 12 = 204$

6\. 384

7\.
$$\begin{array}{rrrr} & 4 & 1 & 7 \\ \times & & & 6 \\ \hline 2 & 4 & 0 & 0 \\ & & 6 & 0 \\ & & 4 & 2 \\ \hline 2 & 5 & 0 & 2 \\ \hline \end{array}$$

8\. $26 \div 2$, $13 \div 1$, 13

9\.

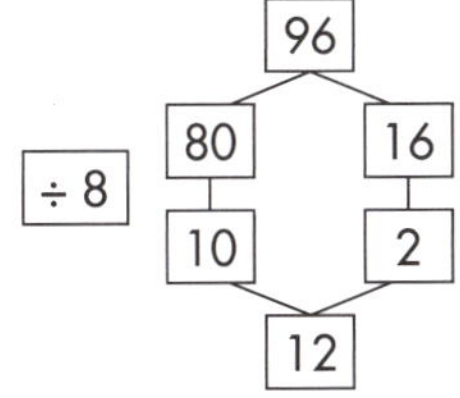

10\. 238

Test 2

1. 28, 32, 48, 56, 84
2. 60, 580, 1000
3. 3456
4. 21 R 30
5. Round 58 to 60, estimate $21 \times 60 = 1260$, actual 1218
6. $6 \times 24 = 6 \times 25 - 6 \times 1 = 150 - 6 = 144$
   $182 \div 7 = 175 \div 7 + 7 \div 7 = 25 + 1 = 26$
7. Sometimes true
8. $72 \div 8 = 9$
   $35 \div 7 = 5$
9. 3407, 2·056
10. 152, 153, 155, 156, 159, 160, 162, 164, 165, 168; total 10

Back to Basics Multiplication and Division Years 5–6

Reprinted 2018

ISBN: 978 1 74215 939 3

Published by Pascal Press
PO Box 250
Glebe NSW 2037
www.pascalpress.com.au
contact@pascalpress.com.au

Author: Ann Baker
Publisher: Lynn Dickinson
Editor: Eliza Hope
Proofreader: Tim Learner
Design and illustration: Janice Bowles
Page layout and technical illustration: Ruth Schultz
Cover design: Deb Snibson, MAPG
Printed by Thumbprints